LINCOLN CHRISTIAN COLLEGE AND SEMINARY

D1404034

The Statistical Exorcist

POPULAR STATISTICS

a series edited by

D. B. Owen
Department of Statistics
Southern Methodist University
Dallas, Texas

Nancy R. Mann
Biomathematics Department
UCLA
Los Angeles, California

Other Volumes in Preparation

The Statistical Exorcist

DISPELLING STATISTICS ANXIETY

Myles Hollander
Frank Proschan

The Florida State University
Tallahassee, Florida

MARCEL DEKKER, INC.   NEW YORK · BASEL · HONG KONG

Library of Congress Cataloging in Publication Data

Hollander, Myles.
   The statistical exorcist.

   (Popular statistics; 3)
   Includes index.
   1. Statistics.   I. Proschan, Frank, [date].
II. Title.   III. Series.
QA276.12.H65 1984   001.4′22   84-17463
ISBN 0-8247-7225-3

Copyright © 1984 by Marcel Dekker, Inc.   All Rights Reserved

Neither this book nor any part may be reproduced or transmitted in any form or by
any means, electronic or mechanical, including photocopying, microfilming, and
recording, or by any information storage and retrieval system, without permission in
writing from the publisher.

MARCEL DEKKER, INC.
270 Madison Avenue, New York, New York 10016

Current printing (last digit):
10 9 8 7

PRINTED IN THE UNITED STATES OF AMERICA

To my wife Glee, whose probability distribution is concentrated on excellence.

To my son Layne, who revises his opinions each spring and wins our baseball pool each fall.

To my son Bart, whose statistical decision making has increased the value of our portfolio by millions (on paper, of course).

M. H.

To my wife Pudge, my daughter Ginnie, and my son Michael—three outliers in a normal family.

F. P.

105682

10563

Preface

Statistics is used in hundreds of ways in everyday life. Hundreds of texts on statistics and its applications are available. Yet only a small proportion of the population understands statistics or how it is used in everyday life. You find these statements hard to believe?

Fifty million people are told the winner of the presidential election after only a relatively small percentage of the votes are counted. Ask them how this minor miracle can be achieved. Millions of young men were told the "random" order in which they would be called up for the draft. How many knew that the procedure used to determine their destinies was *not* random at all because of serious statistical mistakes? Millions of people gamble in luxurious casinos in Las Vegas and Atlantic City, and millions more buy lottery tickets. How many of these gamblers have any idea of their chances of winning? How many still believe that cleverly devised systems of gambling can beat the house? How many have peculiar ideas about a "law of averages"?

Since so many people are affected by statistics in so many basic activities, why does the appalling ignorance of statistics persist in spite of the great number of books on the subject? Simple—the books are written in a *foreign language*. Not literally, but the effect is the same. The books primarily explain the *mechanics of statistics* using the *language of algebra*. This is appropriate if the aim is to teach relatively few people how to *do* statistics. However, it is self-defeating if the aim is to reach a great many people, since most people are afraid of algebraic arguments and cannot follow them.

Our aim is completely different. We aim to narrate in plain English words how statistics is used (and sometimes misused) in everyday life. We have written 26 vignettes which illustrate real-life applications of statistics. But what is crucial is the mode of communication: *we use no algebraic symbols*; rather, we use English supplemented by $+$, $-$, $\times$, and $\div$, the standard symbols of arithmetic.

We believe the book is easy to read and understand. We have also added some features to make the book enjoyable: cartoons and quotations relating to statistics. (We have deliberately cheated in some cases and fabricated quotations from nonexistent books; we challenge the reader to spot these. Readers can check their list against our list in Appendix III, our confession of phony quotations.)

We wish to point out that the book may be useful to a variety of audiences.

1. *Anybody* who wishes to learn about the many uses of statistics in the real world. If they can read this preface, they can read the book.

2. Any student of elementary statistics. Our book is currently being used in a new freshman-level introductory statistics course at Florida State University. The purpose of the course is to teach statistics using examples (rather than using mathematical formulas). Our book may also serve as a useful and stimulating supplement for courses in which the basic text may be more concerned with teaching the student the techniques of statistics.

3. Any applied statistician who knows the techniques of statistics and would like to see how statistics is used in fields other than those he has worked in.

4. The theoretical statistician who knows strange and frightening things about the behavior of esoteric random elements, who spends his days at or near infinity, who deliberately creates functions knowing full well he will soon heartlessly speed them on their way to 0, and who lives in an imaginary world which in his madness he views as characteristic. Even such a mathematical wizard may benefit from reading our book to see how his theory finds a myriad of applications in the real world.

The core of this book consists of 26 vignettes, each describing an application of statistics. The vignettes fall mainly into four categories:

Making decisions
Sampling
Learning from data
Estimating probabilities

The placement of a vignette into a specific category should not be interpreted too rigidly. Many of the vignettes easily fit into several categories.

It may be necessary to calculate or estimate probabilities in order to make decisions; proper sampling procedures enable us to learn from data and make sound conclusions; and so forth. Vignettes are self-contained so readers and teachers can focus on the ones they enjoy and skip others.

In each vignette several words or phrases appear in **boldface** letters. Such terms are particularly important and are informally defined in the context in which they arise.

At the end of each vignette there are a number of problems. The harder problems are signified by an asterisk (*). The problems are fun, stimulate readers to deeper thought, and hopefully arouse their curiosity to learn more about the techniques of statistics. Answers to selected problems are given in Appendix II. It is unfortunately (perhaps fortunately) true that our book does not teach readers exactly what to do to solve a few of the harder problems. They may find themselves stimulated to delve more deeply into other texts on statistical techniques so that they can solve these harder problems. If our book has aroused their passion for statistics, we will not feel jealous if this passion leads readers to spend evenings with other books in statistics.

We are grateful to Glee Hollander and Virginia Proschan Smoliar, who drew cartoons for love rather than for money; to Debbie Pryor, who drew one cartoon for love and a laughable amount of money; to Kathy Collins, Susan Lincicome, Betty Tanner, Cindy Toler, Lena Townsend, and Debbie West, who typed and assisted for love and money; to Dewey Rudd, who did figures for artistic pleasure, to Bart Hollander for assistance in preparation of the index; and to our many colleagues and students who stood by helplessly and passively, neither encouraging nor dissuading us. We thank John Rasp for helpful comments, James Sconing for checking the solutions in Appendix II, Arif Zaman for useful suggestions, and John Kitchin for generating part of Table A of Appendix I.

We warmly thank the Air Force Office of Scientific Research, AFSC, USAF for supporting our research in theoretical and applied areas of statistics. These studies led to applications described in some of the vignettes.

We are indebted to the world for lending itself so neatly to the application of statistics.

Finally, we are grateful to the many standard statistics text writers who have ignored the vast majority of the population who cannot read their texts. By default, they have generously left us a very large audience who may wish to learn about the everyday uses of statistics from a book written in the language they use everyday.

Myles Hollander
Frank Proschan

Contents

Learning from Data

Estimating Probabilities

An Invitation

To the unfortunate students of statistics, who

1. Learned just enough to recognize, fear, and detest statistics
2. Succumbed at a heartbreakingly early age of dread terminal statistics
3. Got through statistics as adroitly as a dog through water, absorbing as little as possible, ear flaps firmly anchored, and shaking off immediately any of the foreign material still adhering
4. Were joined with hostile instructors in a deep mutual abiding hatred, jointly dreading their seemingly interminable hours together
5. Harbored a gnawing suspicion about their instructors' misplaced passion when discussing normal curves
6. Viewed statistics as an intellectual version of sadistics

Here is your chance to learn painlessly the many ways in which statistics is useful. You will enjoy reading, undistracted by symbols, what statistics can do for us.

The Statistical Exorcist

MAKING DECISIONS

"Daniel's a statistician. He sees numbers—fractions, equations, totals—and they spell out the odds for him. God knows he's brilliant at it; he's saved the lives of hundreds with those statistics."

Robert Ludlum, *The Parsifal Mosaic*

"So we've lost $6,000. Now here's my plan: 6, 4, 3, 17, 22, 22, 31."

Vignette 1

Gambling at Roulette

We didn't stop, except for gas, until after midnight, when we had passed the border and were in Monte Carlo. Evelyn insisted on seeing the casino and playing at the roulette table. I didn't feel like gambling, or even watching, and sat at the bar. After a while she came back, smiling and looking smug. She had won five hundred francs and paid my bar bill to celebrate. Whoever would finally marry her would marry a woman with sound nerves.

Irwin Shaw, *Nightwork*

If you are looking to cheat the house or beat it legitimately, forget about roulette. It's basically a sucker's game.

Mario Puzo, *Inside Las Vegas*

"If we make prudent investments, and forget about the trip to Eleuthera, the odds are 7 to 1 that we will triple that inheritance within five years."

Erica was tired of this kind of talk. "Forget about the odds. Let's take some bold steps. What happened to the John Wayne type I fell in love with?"

"Honey, the world is full of uncertainties. Life is gray—not black or white—and we've got to deal with the cloudy areas by taking the path that maximizes our expectations."

Erica pondered the uneventful years of uncertainties with Vance. Not only wasn't she maximizing her expectations, she was probably minimizing her life expectancy. One thing was certain. She would end the uncertainty of living with this probability-calculating computer.

Margo Brooks, *First Journey Home*

Every year millions of people gamble at roulette in the casinos of Nevada, Atlantic City, and elsewhere. The roulette apparatus consists of a table with a built-in wheel and a small steel ball. In the United States the wheel has 38

3

small compartments (called slots or pockets) numbered 0, 00, and 1, 2, 3, ..., 36. (The notation 1, 2, 3, ..., 36 is a shorthand way of writing the complete list of successive integers starting with 1 and ending with 36.) The slots numbered from 1 to 36 are alternately red and black; the slots numbered 0 and 00 are on opposite sides of the wheel and are green. Each number and color on the wheel has a corresponding space on the table. There are a number of possible bets, but for now we will discuss only the bet in which the player places chips on one of the 38 numbers, 1, 2, 3, ..., 36, 0, 00. The croupier spins the wheel, the wheel rotates many times, and finally the ball comes to rest in one of the slots, by chance, thus determining the point for that play.

We assume a balanced or "fair" wheel; that is, each of the 38 possible outcomes 1, 2, 3, ..., 36, 0, 00 is equally likely to occur.

Now if the player places, say, $1 on number 5 and the steel ball comes to rest in slot number 5, he receives from the house $36 (his own $1 back plus $35 winnings from the house). If the ball comes to rest in any slot *other* than slot number 5, the house takes his $1.

A number of questions come to mind.

1.   What is the player's chance of winning on a single bet?
2.   What is the house's chance of winning on a single bet?
3.   If the player bets repeatedly following any scheme he chooses, what percentage of the total amount that he bets will he lose (or win) on the average?
4.   Is there any system of betting that he can follow that will increase his chances of winning?
5.   Suppose the player starts with a fixed sum of money, say $100, and decides to keep playing until he either loses it all or wins some predetermined amount, say $350, and then quits. Is there any one scheme of playing which is preferable to all other possible schemes?
6.   If the odds are against the player (as we shall soon see), how come a couple of young mathematicians beat the house consistently, until they were gently persuaded to leave town?

Answers

1. Since all 38 possible outcomes 1, 2, 3, ..., 36, 0, 00 are equally likely on a fair wheel, the chance is 1 in 38 (that is, 1/38) that the player wins, that is, in the example described that the number 5 he chose turns up.

2. Since the house wins if any other outcome occurs, the house has 37 chances out of 38, that is, a 37/38 chance (or **probability**) of winning.

3. Suppose the player places a $1 bet 380 times in succession; for the moment, we disregard the particular sequence of numbers he picks. On the average, he wins 1/38 of the bets, that is, 10 bets, and loses 37/38 of the bets, that is, 370 bets. Each of the 10 winning bets yields him a winnings of $35, so that the total winnings he receives from the house is $350. On the other hand, he loses, on the average, 37/38 of the 380 bets placed (that is, he loses a total of $370). Thus, his net loss in 380 games (on the average) is $20. In the long run, he is losing $20 out of every $380 bet, or at the rate of (20/380)100% = 5.3% of the money he bets.

There are many subtle points hidden in what we are saying. For example, we do not claim that in any one sequence of 380 bets of $1 each, the player will lose exactly $20. In fact, in any such sequence, he may *lose* every bet, so that he loses $380 (the chance of this occurring is $(37/38)^{380\dagger} = .00004$, a pretty small chance). Alternatively, he may *win* every one of the 380 bets, so that he ends up $380 \times $35 = $13,300$ ahead (the chance of this occurring is $(1/38)^{380}$, which is about 5 chances in the number consisting of 1 followed by 601 0s, an incredibly small chance.) What we are saying is that in the long run, he will lose about 5.3% of the total of the bets placed.

(Don't feel badly if you are not following any of these calculations. The next vignette is an exact duplication of the present vignette. Some people learn on the second try what they cannot absorb the first time around.)

Of course, the question that naturally occurs to the reader is, What do you mean by "in the long run"? Do you mean 380 bets (as in the example just presented), or 3,800 bets, or 38,000 bets, or just how many bets? To give a precise answer would require probability theory far beyond the scope of this type of book. Rather, we will content ourselves with the relatively crude statement that the greater the number of bets placed, the more likely it is that the percentage of bets lost by the player is close to 5.3%.

4. We pointed out in discussing question 3 that the player's long-term percentage loss in placing successive bets in an arbitrary manner is 5.3%. In an effort to convert this percentage loss to a gain, players have proposed a variety of systems of play that are alleged to beat the house. Unfortunately,

---

$\dagger (37/38)^{380}$ is a shorthand way of writing the product of 380 factors, each of which is 37/38. That is,

$$\left(\frac{37}{38} \right)^{380} = \underbrace{\frac{37}{38} \times \frac{37}{38} \times \cdots \times \frac{37}{38}}.$$

The factor $\frac{37}{38}$ appears 380 times.

statistical theory shows that *no system can beat the house.* An even stronger statement can be made: All systems yield equal long-term percentage losses, namely, 5.3%. Specifically, no matter what sequence of numbers are chosen in successive bets, or what sequence of amounts are bet, the long-term percentage loss remains unchanged; that is, the player loses 5.3% of the total amount bet.

The key ideas behind this unhappy, disillusioning fact are that (1) each of the 38 outcomes (1, 2, 3, . . . , 36, 0, 00) is equally likely, namely, each has a 1/38 chance of occurrence; and (2) the outcome in one spin of the wheel does not depend on the outcomes of previous spins—that is, the wheel is "memoryless." Thus, systems which rely on selecting the present point based on the history of outcomes previously observed are doomed to do no better (or worse) *on the average* than systems which select the present point at random or by whim. Putting it simply, *all systems are equally good* (actually, equally bad) in playing against the house *in their long-run performances.*

This applies not only to systems in which the successive points are chosen in some special way, but also to systems in which the *size* of the bet is chosen in some particular fashion, usually based on past outcomes of the points or the past history of wins and losses. In fact, the house loves to see a player with a system take his place at the roulette wheel. They *know* that such a player will persist in his play until his last dollar is gone. A less "scientific" player might leave with some of his money if he were to experience a succession of losses.

5. The player starts with a fixed sum of money, say $100, and wishes to either lose his entire stake or win some predetermined amount, say $350, and then quit. What is his best strategy of play? The answer is surprisingly simple. He should play *boldly.* Initially, he places the bet ($10) necessary to achieve his goal of winning $350 in one play. If he wins, he walks away with his goal achieved. If he loses, he places as his second bet the amount necessary to recoup his $10 loss *and* to win an additional $350. (Small practical difficulties crop up in carrying out the policy described, since the house will not accept bets that are not multiples of $1; a compromise is to round up to the nearest dollar. In the present case, the compromise bet would be $11). In general, at every stage the player bets the amount needed to recoup his losses and win an additional $350; if not an integer, the bet is rounded *up* to the nearest integer. Thus, if he is losing consistently, he keeps increasing his successive bets. If at a later stage, he has insufficient money left to reach his goal (of achieving a net gain of $350) on a single bet, he simply bets all that he does have left.

In summary, *he bets as much as necessary to achieve his goal on the current bet.* The reasoning behind the strategy described is that by following this strategy he will be betting on the average, a *total* amount, considering all the bets placed, that is no greater than the total amount he would have bet under any alternative strategy. Intuitively, since the game is unfair to the player (that is, he wins a smaller amount than the odds indicate he should), he should follow that strategy which will result in the smallest total amount bet, on the average. The strategy just described achieves this goal. Of course, the fact that the policy described is actually the best possible can be proved mathematically; the proof requires a knowledge of probability beyond the scope of this book.

6. Two young mathematicians actually did beat the house *consistently* at roulette. (Naturally, the casino did not accept such a beating gracefully. The two young mathematicians were politely but firmly discouraged from further roulette play; to remain in good health, they were encouraged to return home. At the same time, the casino took countermeasures (to be explained subsequently) to prevent the recurrence of such unfortunate losses.) If our claim that the house pays odds of 35 to 1 instead of the "fair" odds of 37 to 1 is correct, how was this possible?

Actually, no inconsistency exists. Our calculation of the odds of winning at roulette was based on the assumption that each of the outcomes 1, 2, 3, . . . , 36, 0, 00 is equally likely. This assumption holds for an "ideal" roulette wheel. However, as the wheel ages, the repeated spins gradually wear the wheel unevenly so that certain points become more likely to appear than other points. Thus, the chance of, say, point 7 appearing is no longer 1/38 but may now be 1/20.

The two young mathematicians had simply observed the outcomes for a specific roulette wheel during the course of thousands of plays, and noted the relative frequency of appearance of each of the 38 points. They then repeatedly placed their bets on the point or points among 1, 2, 3, . . . , 36, 0, 00 that appeared distinctly more frequently than 1 time in 38. Thus if, for example, point 7 did appear, on the average, 1 time in 20, and they were repeatedly betting, say $1, on point 7, they would receive a winning of $35 1 time in 20 and lose $1 on each of the remaining 19 bets. Thus, on the average, they would be winning at the rate of $16 per 20 plays—a quite respectable net gain of 80% per $1 bet.

When the casino realized that the two young mathematicians were taking advantage of the biased behavior of the wheel, they reacted initially as described; namely, their formidably persuasive spokesmen suggested that the young men discontinue their play and return home in good health. However, in addition, they took more positive measures to prevent other enterprising players from taking advantage of any bias that might develop in their roulette wheels. They simply exchanged the physical positions of (say) 5 roulette wheels under the tightest possible security. This was done often enough that no player could build up enough experience with any given roulette wheel to take advantage of its bias.

Ethier (1982) highlights some famous successes at the roulette tables. We quote from her article.

> Near the end of the 19th century, a British mechanic named Jaggers won the equivalent of about 65,000 pounds at the Monte Carlo casino (Kingston 1925, Ch. XV). In 1947, Albert Hibbs and Roy Walford, students at the University of Chicago, parlayed $100 into $14,500 before quitting $6,500 ahead in Reno (*Life,* Dec. 8, 1947, p. 46). In 1950–1951, a 20-member gambling syndicate raked in earnings estimated as high as 6,000,000 pesos (more than $420,000) at the government-owned casino in Mar del Plata, Argentina (*Time*, Feb. 12, 1951, p. 34). . . . Each of these successes was achieved at the roulette tables by betting on the numbers with the highest observed frequencies.

Remark: Odds. Sometimes probabilities are stated in terms of **odds**. (Recall the answer to question 6.) For example, we learned that if a player bets on a specific slot (say slot number 2) at roulette, his probability of winning is 1/38, and his probability of losing is 37/38. We could restate these probabilities as follows:

The odds are 37 to 1 against winning.

Equivalently, we could say,

The odds are 1 to 37 in favor of winning.

In other words, the **odds against** an event is the ratio of the probability that the event does not occur to the probability that the event does occur. Thus

$$\text{Odds against winning} = \frac{\text{probability of not winning}}{\text{probability of winning}} = \frac{37/38}{1/38} = \frac{37}{1},$$

or 37 to 1 (sometimes written 37:1).

The **odds in favor** of an event is the ratio of the probability that the event does occur to the probability that it does not occur. In the roulette example where we bet on slot number 2,

$$\text{Odds in favor of winning} = \frac{1/38}{37/38} = \frac{1}{37},$$

or 1 to 37 (alternatively written 1:37).

Summary

In betting \$1 against the house in the game of roulette, assuming all 38 outcomes (1, 2, 3, . . . , 36, 0, 00) are equally likely, the player has 1 chance in 38 of winning \$35, whereas the house has 37 chances out of 38 of winning the player's \$1. In the long run, this yields the house a $[(37/38) - (35/38)]100\% = 5.3\%$ net gain of all bets placed; correspondingly, the player loses 5.3% of the total money he bets in the long run. No system of play can decrease the player's long-run percentage net loss if the roulette wheel is balanced (that is, all 38 outcomes are equally likely).

If the player sets as his goal to win a specified amount or lose completely his initial capital, his best method of play is to bet initially the wager needed to win the amount he has specified. If he wins, he has achieved his goal; if he loses, he bets the amount needed to recoup his loss *and* win in addition the amount he has specified. He continues in this fashion, always trying to achieve his goal on the *current* bet. He stops, of course, when he has either achieved his goal or has lost all his money. At every stage, he never bets more than the amount needed to just achieve his basic goal (subject, of course, to the house restriction of betting in multiples of \$1). This method is best, simply because the *total* amount bet tends to be smaller than under competing strategies.

These conclusions apply *when the wheel is balanced*. If the wheel wears after repeated play so that one or more of the points 1, 2, 3, . . . , 36, 0, 00 occurs more frequently, on the average, than 1/38, the player can actually beat the house by betting on such point(s).

Another caveat: these conclusions apply to *long-run* play. In any individual play or sequence of plays, deviations may occur due to chance.

Problems

1. The player at the roulette wheel may bet against the house in a second fashion. Half of the numbers among 1, 2, 3, . . . , 36 are colored black, while the remaining half are colored red; the two house numbers 0 and 00 are colored green. The player places his bet, say \$1, on red or black. If the ball lands in a slot of the color he chooses, he wins \$1 from the house; if the ball lands in a slot of the opposite color, or in 0 or 00, he loses his \$1 bet. Assume that all points 1, 2, 3, . . . , 36, 0, 00 are equally likely.
   a. What is the chance that the player: (1) Wins the bet? (2) Loses the bet?
   b. If the player repeatedly bets \$1, choosing black on some bets, red on the remaining bets, what is the: (1) Long-run percentage loss of

the player? (2) Long-run percentage gain of the house? *Hint:* Determine, on the average out of each 38 plays, how many times the player wins $1 and how many times he loses $1. The net loss divided by 38 represents the long-run average loss per game; if you now multiply by 100, you have computed the long-run percentage loss.

c. Is there any system of betting that is superior as far as long-run percentage loss is concerned? Justify your answer.

d. Suppose the player starts with an initial capital of $100 and wishes to play until he either wins $100 or is wiped out. What system of betting should he follow so as to maximize his chance of winning the $100? What is the chance that he *does* win the $100 following this best system? Explain why the system you have proposed is better than competing systems.

e. What odds should the house give to make the bet a **fair** bet; that is, such that the amount won by the house is approximately the same as the amount lost in the long run.

2. a. In the type of roulette bet described in the text (betting on one of the numbers 1, 2, 3, . . . , 36, 0, 00), what odds should the house give to make the bet a fair bet?

b. Suppose in a single game the player places a separate $1 bet on each of the numbers 1, 2, 3, . . . , 36, 0, 00. What is the chance that the player wins more money than he loses? The answer to this question reveals quite clearly the advantage that the house has.

3. The gambling casino will graciously accommodate a group of, say, four or five relatively affluent players who wish to play a private game of poker. For furnishing the luxurious private room, a professional dealer, free drinks served by beautiful women (on the average), protection against cheating or stickups, and so forth, the house modestly and unobtrusively takes a small percentage of each "pot" (the total amount at stake in a single game). If the house takes 5% of each pot, the average pot is $1,000, and an evening's entertainment consists of 100 games of poker, how much will the house collect for its hospitality? If the conviviality extends far into the night, is it possible that not one player emerges ahead?

4.* Dubins and Savage (1965) consider the "red-and-black" type of betting as described in Problem 1. To be specific, suppose a gambler with an initial fortune of $1,000 attempts to attain $10,000 by playing the red-and-black style. Give a heuristic argument to show that the chance of achieving this goal by any strategy is less than 1/10. (Dubins and Savage show that the chance of achieving this goal using bold play is .082.)

VIRGINIA SMOLIAR

"Professor, what's the good of winning all that money, if you have to spend it paying hospital bills?"

"Yeah doc, it would be smart to leave town before you have an accident, maybe fatal."

5. The Dubins and Savage (1965) book is an advanced mathematical work on how to gamble optimally when playing unfavorable games. Breiman (1961), in an advanced paper, considers optimal gambling systems for favorable games. Look at these two references. Now do you appreciate our nonsymbolic approach?

6. Not all games in the casino are unfavorable to the player. The card game blackjack can actually be favorable to the player if he plays correctly. Thorp (1962) presents strategies that enable the player to gain a steady advantage over the house. Read Thorp's book, play Thorp's system in Las Vegas, and then ask a friend to write your response to the following question: Did anyone (try to) break your knuckles?

7. What are some real-world situations in which the gambler is allowed to bet more than he possesses?

8. The gambler of Problem 4 wants to go from $1,000 to $10,000 playing

red and black at roulette. Suppose he plays the very timid strategy of betting $1 on each play. Explain why this is a very poor strategy.

9. What are some nonmonetary advantages of playing the strategy described in Problem 8?

10. A disillusioned and self-destructive gambler wants to lose his fortune of $1,000 at roulette in as few plays as possible. How should he gamble?

11. Roulette at Monte Carlo is slightly different from roulette in the United States. At Monte Carlo, the wheel has slots labeled 1, 2, 3, . . . , 36 and an additional house slot labeled 0; however, there is no slot labeled 00. Half of the numbers 1, . . . , 36 are red, and half are black. The slot 0 is green. What is the chance of a win if you bet on red? Why do Las Vegas casinos make more money at roulette than do Monte Carlo casinos?

References

1. L. Breiman (1961). Optimal gambling systems for favorable games. *Proceedings of the Fourth Berkeley Symposium on Mathematical Statistics and Probability 1*, 65–78. University of California Press, Berkeley, California.

2. L. E. Dubins and L. J. Savage (1965). *How to Gamble If You Must.* McGraw-Hill, New York.

3. S. N. Ethier (1982). Testing for favorable numbers on a roulette wheel. *Journal of the American Statistical Association 77*, 660–665.

4. C. Kingston (1925). *The Romance of Monte Carlo.* John Lane, The Bodley Head, Ltd., London.

5. E. O. Thorp (1962). *Beat the Dealer.* Blaisdell, New York.

References 1 and 2 use advanced mathematics and are *not* suggested for the nonmathematically trained reader. Reference 3 is a technical paper which addresses the problem of deciding, on the basis of observed frequencies of the various numbers on a given roulette wheel, whether that wheel has any favorable numbers (that is, whether any of the numbers 1, 2, . . . , 36, 0, 00 has a probability of occurrence that exceeds 1/38). Reference 4 describes the pleasures of Monte Carlo, circa 1925. Reference 5 is very readable and entertaining, and does not require knowledge of mathematics.

Vignette 2

The Dowry Problem

Probably the women had already cast about among the families of the men who might now be called his friends, for that prospective bride whose dowry might complete the shape and substance of that respectability Miss Coldfield anyway believed to be his aim.

William Faulkner, *Absalom, Absalom!*

To display the greatness of the dowry, Mtalba began to put on some of the robes to the accompaniment of a xylophone made of bones played by one of her party, a fellow with a big knobby ring on his knuckle.

Saul Bellow, *Henderson The Rain King*

And though the question of dowry never arose between our families, I would not like to lose sight of this ancient and honored custom. For in what other manner can a man show his affection for his daughter and appreciation of her husband?

Harold Robbins, *The Pirate*

E'en a Hunchback like Mr. Pope had greater Opportunity than a Lass with a straight Back, Quick Wit, but no Dowery!

Erica Jong, *Fanny*

But what husband could I, without hope of dowry, ever find? Certainly not a generous Captain who outfits strange ladies like queens.

Salman Rushdie, *Shame*

In India, as many as 400 young married women are killed each year by their husbands and in-laws for failing to provide an adequate dowry.

Among Hindus, it is customary for the bride's parents to pay the groom's family a dowry. And it is common for her husband and in-laws to extort higher payments from her family, even after the amount has been set. If they balk or fall behind on the payments, the bride may be harassed and even slain. . . .

In recent years, Indian feminist groups have lobbied for thorough investigations of dowry deaths, with harsh penalties for offenders. For the most part, however, dowry abuse and bride burning go unpunished.

From an article "Scandal in India" by Irving Wallace, David Wallechinsky, and Amy Wallace. Based on a suggestion by Linda D. Banford. *Parade Magazine*, January 8, 1984

A bachelor is given the (enviable?) task of trying to choose the girl with the largest dowry among four girls. (A dowry is, of course, the money, goods, or estate that a woman brings to her husband in marriage. It is common practice in India, for example, but it is not a custom in the United States.) Rather than being told the values of the four dowries, he is presented in random order four slips of paper, each with a different dowry written on it. When he is presented a slip, he must either choose it or reject it. Once rejected, the slip cannot be chosen later. If the bachelor chooses the slip with the *largest* dowry, he gets the dowry and (fortunately or unfortunately) the girl corresponding to that dowry; *otherwise he gets nothing!* We assume that the bachelor does not know anything about the distribution of possible dowries. What is the optimal strategy for the bachelor? That is, how should he make his choice so as to maximize his chance of selecting the largest dowry? (If he selects the largest dowry, we say he has *won* the dowry "game.")

A much more mundane version of the problem may point up its essential statistical features. A game is to be played in which the player is to select the tag having the highest number among four numbered tags placed in an urn.[†] The player draws a tag at random. After drawing the first tag, he must either accept it, ending the game with a win if it so happens that the first one bears the largest number, or losing if he accepts the first one and it does not bear the largest number. If he does not accept the first tag, it is *permanently rejected,* and the player picks another tag at random out of the remaining three. The process continues: at each stage the player must either accept the tag he has just selected, ending the game, or he must permanently reject it. [Of course, if he rejects the first three, his choice must be the

---

[†] I don't believe I've ever seen an urn except in museums. In literature, they almost always contain some person's ashes. Maybe probability would be more popular and less frightening, if statisticians shied away from the creepy language they use with high probability. F. P.

"I'm sure the book strategy for winning the largest dowry is correct. But do I want to win?"

remaining (fourth) tag in the urn.] The question is, What strategy gives the highest chance of winning?

The assumption that the player knows nothing about the distribution of the numbers on the tags means that, after each tag is drawn, the player knows only the rank of the number on the tag among the numbers drawn so far. (The rank of the smallest number is 1, the rank of the second smallest is 2, and so on.)

We say that a draw is a *candidate for selection* (or simply a *candidate*) if the number drawn exceeds those of all earlier draws. It can be proved that the best strategy in the urn game is: *The first draw is rejected no matter what number appears. Then, following this rejection, accept the first candidate drawn, if any.*

The best strategy in the dowry problem is also of this form. This, of course, is true since the urn game and the dowry problem have the same mathematical structure.

In Table 2.1, we list the 24 possible permutations when one is drawing four tags from the urn. Thus, the notation 3241, for example, means that the tag with the third smallest number is drawn first, the tag with the second smallest number is drawn second, the tag with the fourth smallest (that is, largest) number is drawn third, and the smallest number is drawn last. We

Table 2.1   Wins under Each of Two Strategies

1234	2134	3124* [†]	4123
1243[†]	2143* [†]	3142* [†]	4132
1324[†]	2314[†]	3214* [†]	4213
1342[†]	2341[†]	3241* [†]	4231
1423*	2413*	3412*	4312
1432*	2431*	3421*	4321

*Wins by accepting the first candidate (if any) after rejecting the first number drawn.
[†] Wins by accepting the first candidate (if any) after rejecting the first two numbers drawn.

use an asterisk (*) to mark those permutations which lead to a win when the player uses the strategy reject the first number drawn, then accept the first candidate, if any. Thus, the permutation 1234 does *not* lead to a win with this strategy, because after the first draw of 1, the subsequent draw 2 is a candidate, and thus according to the strategy it is accepted; the player loses because the largest number is *not* selected. On the other hand, the permutation 1423 leads to a win because after the first draw of 1, the next draw 4 is a candidate, and thus the strategy dictates that it be accepted.

We have also marked with a dagger ([†]) the permutations which yield a win by the strategy that rejects the first two numbers drawn, then accepts the first candidate, if any.

Since the tags are drawn at random, each of the 24 permutations is equally likely to occur. Thus, each permutation has chance 1/24, and we can easily calculate the probability of winning with various strategies. For example, the strategy, accept the first candidate (if any) after rejecting the first two numbers drawn, wins for 10 of the 24 permutations, and thus has a chance of winning equal to 10/24. The reader should be able to directly verify the chances of winning given in Table 2.2.

The dowry problem and the urn game can be formulated and solved for any number of dowries (or tags). The *form* of the best strategy remains the same: *After a specified number of successive dowries are automatically rejected, pick the first candidate (if any).*

The specified number initially rejected depends solely on the number of competing dowries. By making up a table similar to Table 2.1, you can readily determine this specified number.

The basic dowry problem can of course be rephrased in different settings and has many interesting generalizations. Our approach here is based on the scholarly article by Gilbert and Mosteller (1966) to which the reader can turn for further information.

Table 2.2 Chances of Winning with Various Strategies

Strategy	Chance of Winning
Accept the first number drawn	6/24
Accept the first candidate (if any) after rejecting the first number drawn	11/24
Accept the first candidate (if any) after rejecting the first two numbers drawn	10/24
Accept the fourth (last) number drawn	6/24

Summary

In some precisely formulated models, it may be possible to determine the best strategy to follow. The best strategy does not necessarily guarantee complete success; it simply increases the chances of success as compared with competing strategies.

We have illustrated the way to determine the best strategy in a somewhat artificial problem—the well-known dowry problem. The mathematical reasoning is correct; the weakness lies in the fact that a complex human situation has been reduced to a model describing a few quantitative factors. Life just isn't that simple—thank God!

Problems

1. Suppose there are three dowries. What is the best strategy, and what is the chance of selecting the largest dowry following it?
2. Suppose there are five dowries. What is the best strategy, and what is the chance of selecting the largest dowry following it?
3. Rephrase the dowry problem in the context of an executive attempting to hire the best secretary. In particular, what could play the role of the dowry?
4. Rephrase the dowry problem in the context of a boy attempting to choose the prettiest of several unseen and unknown girls.
5. What is the chance of selecting the largest dowry with a strategy of *random* acceptance when there are four dowries?
6. Consider the following four competing strategies for the case when there are four dowries.
   I: Accept the first number.
   II: Accept the second number.

III: Accept the third number.

IV: Accept the fourth number.

The chance of selecting the largest dowry is the same for each of the four strategies. Explain.

7.*   Consider the following extension of the dowry problem. Numbers are presented to the player searching for the largest in an order that is determined by a hostile opponent, that is, an opponent who wishes to minimize the player's chance of finding the largest among four numbers. After each draw, the rank of the new number among those so far drawn is given to the player. At this point the player must reject the new number or choose the new number. Can the opponent reduce the player's chance of selecting the largest dowry to 1/4? Explain.

8.*   Refer to Problem 7. Why can't the opponent reduce the probability that the player will select the largest dowry to a value less than 1/4?

9.   Reformulate Problems 7 and 8 (and provide appropriate solutions) when there are five numbers instead of four numbers.

10.   Give a real-life example of the situation described in Problem 7.

Reference

1.   J. P. Gilbert and F. Mosteller (1966). Recognizing the maximum of a sequence. *Journal of the American Statistical Association 61*, 37–73.

Although there are many generalizations, spinoffs, and variations of the dowry problem, reference 1 gives a relatively comprehensive treatment of the subject. Some parts of the paper require advanced mathematics, but many parts can be enjoyed by the nonmathematical reader (who can ignore some of the mathematical derivations).

Vignette 3

A Tie Is Like Kissing Your Sister

"I just couldn't have lived with going for a tie," said Bowden. "Don't think my team could have either." [Comment by Florida State University's football coach Bobby Bowden after FSU, trailing the University of Miami 10-3, scored with 39 seconds left in the game, cutting the deficit to 10-9. FSU attempted a two-point play to win the game; the two-point attempt was unsuccessful, and Miami won 10-9, ending FSU's 18-game regular season winning streak.]

Tallahassee Democrat, September 29, 1980

"We're trying to win the game," Osborne said Monday night. "I don't think we should go for a tie in that case. We wanted an undefeated season and a clear-cut national championship." [Comment by University of Nebraska's football coach Tom Osborne after the 1984 Orange Bowl game. Osborne was defending his decision to go for two points after Nebraska had scored with 48 seconds left in the game, cutting the University of Miami's lead to 31-30. The two-point attempt failed, and Nebraska lost 31-30. Miami's victory secured the national championship. Nebraska's final ranking was number two in the nation.]

Gainesville Sun, January 4, 1984

"I would have gone for one point and the national championship," said Dodd. "I'm fond of Tom Osborne but I'm going to tell you what he should have done.

About 10 days before the game—when everybody was quiet and sane— he should have called his squad together and said 'now listen, men, if this game should come down to one point to tie and two to win, let me tell you what will happen.

If you go for two the odds are about 5-to-1 against your making it against a good football team. If you go for one, the odds are about 19-out-of-20 you will make it. If you go for two and fail, you've lost the Orange Bowl—and the national championship. If you go for one and tie, you win the national championship.

Now what do you want to do?'"

19

[Comment by former Georgia Tech coach Bobby Dodd, as quoted in an article by Bill McGrotha, *Tallahassee Democrat,* January 12, 1984. McGrotha goes on to say "Clearly, the way Dodd would have sold it, the team would have voted for the one point in that circumstance."]

"I have to make a scene to get her out of the house for a little football," said Humboldt. She threw a very good pass—a hard, accurate spiral. Her voice trailed as she ran barelegged and made the catch on her breast. The ball in flight wagged like a duck's tail. It flew under the maples, over the clothesline. After confinement in the car, and in my interview clothes, I was glad to play.

Saul Bellow, *Humboldt's Gift*

They made a picture—two tough old guys who probably had no wish to add to each other's troubles, positively had no taste for losing, and wouldn't admit how tickled they both were with the tie.

Red Smith, *To Absent Friends*

Suppose a college football team is trailing its opponent by seven points in the closing moments of the game, and it scores a touchdown, thereby reducing its deficit to one point. The coach of the team now must make a choice between two options. The team can attempt a two-point play via a run or a pass, winning the game if it succeeds and losing if it fails. Alternatively, the team can attempt to kick an extra point, tying the game if the kick is good and losing if the kick misses. Although all sane coaches prefer a win to a tie, it is harder to make a two-point play than to make the relatively easy one-point kick; thus, the choice of strategy is sometimes a difficult one for a coach to make. The selection of a strategy should be made on the basis of the following three factors:

1. The chance of successfully making a two-point play if one is attempted
2. The chance of successfully kicking for an extra point if attempted
3. The relative value (utility, say) of winning the game versus the relative value (utility) of only tying the game

Suppose that the team's chance of making a two-point play is 1/2 and that its chance of kicking an extra point is 9/10. Suppose further that the coach feels a win is twice as good as a tie. We quantify this by saying that the utility of a win is 1, whereas the utility of a tie is only 1/2. We can now ask what is the expected amount of utility to be gained from each strategy? Naturally we prefer the strategy with the larger expected utility. Assuming a loss is worth nothing (has a utility of zero), the expected utility for the two-point strategy is

(utility for winning) $\times$ (chance of making two-point play)
    + (utility for losing) $\times$ (chance of missing two-point play)

$$= 1 \times \frac{1}{2} + 0 \times \frac{1}{2} = .5.$$

Similarly, the expected utility for the one-point strategy is

(utility for tying) $\times$ (chance of making kick)
    + (utility for losing) $\times$ (chance of missing kick)

$$= \frac{1}{2} \times \frac{9}{10} + 0 \times \frac{1}{10} = \frac{9}{20} = .45.$$

Since the expected utility of .50 for the two-point strategy exceeds the corresponding value of .45 for the one-point strategy, the conclusion is that the coach should prefer the two-point strategy.

Of course the chance of making a successful two-point play, the chance of making a successful kick, and the coach's degree of preference for a win over a tie, will all vary from team to team. Above, we took the values 1/2 (chance of making a two-point play), 9/10 (chance of making a one-point play), and the relative utilities of a win to a tie (win twice as valuable as a tie) simply for the purpose of illustration. It is interesting to see how changing these values affects the choice of preferred stategy.

Let us then consider the desirability of each strategy in a slightly more general setting. From the previous reasoning, we see that the expected utility for the two-point strategy is equal to

(utility for winning) $\times$ (chance of making two-point play),

and the expected utility for the one-point strategy is equal to

(utility for tying) $\times$ (chance of making kick).

Thus, the two-point attempt is to be preferred to the one-point attempt if

(utility for winning) $\times$ (chance of making two-point play) is greater than
(utility for tying) $\times$ (chance of making kick).   (1)

Dividing both sides of the inequality by the value "utility for winning" yields the equivalent criterion:

Prefer the two-point attempt if
    (chance of making two-point play) is greater than

$$(\text{chance of making kick}) \times \frac{(\text{utility for tying})}{(\text{utility for winning})}, \qquad (2)$$

and prefer the one-point attempt if the inequality is reversed.
Inequality (2) is relevant to the following five comments:

Comment 1. Recall that in our numerical example the utility for tying is
1/2, the utility for winning is 1, and so the ratio:

$$\frac{\text{utility for tying}}{\text{utility for winning}} = \frac{1/2}{1} = \frac{1}{2}.$$

If we specified instead that the utility for tying is 1 and the utility for winning
is 2, we again find that the ratio:

$$\frac{\text{utility for tying}}{\text{utility for winning}} = \frac{1}{2}.$$

The point is that in deciding which strategy is to be preferred only the *ratio* of
utilities rather than the separate individual values is relevant.

Comment 2. In our numerical example the left-hand side of inequality (2)
is .5, and the right-hand side reduces to .9 × 1/2 or .45. Thus, the two-point
strategy is preferred. However, if the team's chance of making a two-point
play is only .40, say, then the left-hand side of inequality (2) is .40, the right-
hand side is .45, and the one-point strategy is preferred.

Comment 3. One famous football cliché says "A tie is like kissing your
sister," the implication being that there isn't much pleasure to be derived
from a tie. A coach who feels this way would no doubt assign a very low
utility to the prospect of a tie. As an extreme case, suppose he assigned the
value 0 as his utility for a tie. This would be equivalent to believing that a tie
is as bad as a loss, and very few coaches believe that. However, for such an
extreme-thinking coach, the value of the right-hand side of inequality (2)
would be 0. Thus, the two-point strategy would always be preferred because
for any respectable football team the chance of making a two-point play is
greater than 0.

Comment 4. Not all football coaches disdain a tie. To a great extent, the
utility of a tie depends on the circumstances of the game, the team's season
record, and the prestige of the opposing team, to name just a few factors. For
example, Bill Peterson, coach of the 1967 Florida State University
Seminoles, elected late in the game to make a one-point attempt rather than a
two-point attempt with his team down 36–37 to the number one-ranked
college team, the University of Alabama. The extra-point attempt was
successful, producing a memorable 37–37 tie. After the game Peterson
justified his decision by stating that he didn't want to waste the great effort his

"We couldn't go for a tie because I don't have a sister."

players had made in the game. In that situation, Coach Peterson had a relatively high utility for a tie.

Comment 5. Suppose, in our numerical example, the chance of a successful two-point play is .5, the chance of a successful one-point play is .9, the utility for a win is 1, but the utility for a tie is increased to 3/4. Then the left-hand side of inequality (2) is .5, but the right-hand side of (2) is .9 × (3/4)/1 = .675 which is greater than .5. In this situation, the one-point strategy is to be preferred to the two-point strategy.

Summary

1. In the situation described in the vignette, the choice of strategy depends on several factors, namely

a. The chance of successfully making a two-point play if one is attempted
b. The chance of successfully kicking for an extra point if attempted
c. The relative value of winning the game versus the relative value of only tying the game

2. The best strategy for one football team may not be the best strategy for another football team.
3. Summary point 2 obviously extends beyond the realm of football to many other situations. For example, the decision of an individual to make either investment A or investment B will depend on a number of factors, and the best decision may differ from individual to individual.
4. Just because the strategy you used failed, this does not mean that you chose the wrong strategy. Even best strategies do not always succeed.

Problems

1. Suppose the chance of successfully making a two-point play is 6/10, the chance of successfully making a one-point play is 8/10, and the coach's utility for a win is 1 while his utility for a tie is .7. What is the preferred strategy?
2. Suppose the chance of making a successful two-point play is .7 and the chance of making a successful one-point play is .9. What value of the ratio:

$$\frac{\text{utility for tying}}{\text{utility for winning}}$$

leads to indifference between the two strategies?

3.* Consider your favorite college football team. How could you estimate its chance of successfully making a two-point play? How could you estimate its chance of successfully making a one-point play? How could you get the coach to assign a value to the ratio (utility for tying)/(utility for winning)?

4.* The subject of extra-point strategy in football is nicely discussed by Porter (1967). In addition to the situation described here, Porter also analyzes the more complicated situation where late in the fourth quarter, say, a team scores a touchdown to cut its deficit to eight points (say the score is now 6-14), and there is still time left in the game to score another touchdown. Assuming the opponent will not score again, there are three strategies to consider.

Strategy A. Make an extra-point attempt after the first touchdown, with the idea that if it is successful you will again attempt the extra point after the second touchdown, thereby playing for a tie. If the first kick fails, you will make the two-point attempt later for a tie.

Strategy B. Make the two-point attempt after the first touchdown with the idea that if it is successful you will kick later for a win. If the first two-point attempt fails, you will then make another two-point attempt later in order to tie.

Strategy C. Try an extra-point attempt after the first touchdown. If it is successful, a two-point attempt for a win is planned for after the second touchdown. If the kick is unsuccessful, a two-point attempt for a tie is planned for after the second touchdown.

Compare these three strategies on the basis of expected utilities.

5. In the January 1, 1979 Cotton Bowl game between the University of Houston and Notre Dame University, the "Fighting Irish" of Notre Dame were trailing 34-20 late in the game when they scored. They elected to go for two points and succeeded, thereby making the score 34-28. A few minutes later, with just 2 seconds left in the game, Notre Dame again scored and then successfully kicked the extra point to win the game 35-34. (It was a spectacular comeback; at one point Notre Dame trailed by 34-12. God must be a Notre Dame fan!) Does this prove that in Problem 4, strategy B is better than strategy C? Explain.

6. Suppose a college football team is trailing its opponent in the *early* stages of the game, and it scores a touchdown, reducing its deficit to one point. In this situation, what factors affect the selection of strategy, one-point kick or two-point play? Is the analysis described in this vignette still applicable? Explain.

7. Suppose on a given Saturday in September, a college football team, trailing its opponent by seven points in the closing moments of the game, scores a touchdown, elects to try the two-point play, the play fails, and the team loses the game by one point. Should the team's coach change his relative value of winning versus tying in the next game? Explain.

8. Return to Problem 7. Should the coach require that his team devote a significant amount of extra practice time to two-point plays? Explain.

9.* Describe another situation (not relating to football) where choice of strategy is important. Perform an analysis similar to the one in this vignette.

10. Describe a situation in which the relative value of each possible outcome depends solely on the monetary profit to be gained from the outcome.

Reference

1. R. C. Porter (1967). Extra-point strategy in football. *The American Statistician*
    21, 14-15.

Reference 1 does not require advanced mathematics and, even though symbols are used, most readers will find this paper enjoyable and easy to follow.

Vignette 4

Monitoring a Nuclear Reactor

MIDDLETOWN, PA.—Radiation alarms sounded just four minutes after authorities began releasing radioactive krypton gas from the Three Mile Island nuclear power plant Saturday. . . .

Tests showed no excess radiation was released in the 8 A.M. incident. However, officials ordered tests Saturday to work out problems with radiation detectors. . . .

Howard Denton, of the Nuclear Regulatory Commission, said the problem occurred because the initial rush of krypton through the monitors caused an overly high reading before the flow stabilized and the monitor settled to a true reading.

From an AP news release, June 29, 1980

One of the dangers in operating a nuclear reactor is that it may give off dangerously high levels of nuclear radiation if it is not operating properly. To alert the workers, a warning device is installed which gives off a clear-cut signal that the level of radiation is too high when this is the case.

There is a serious difficulty, however, with this safety system. Suppose the warning device malfunctions. Actually it may malfunction in either of *two* ways.

1. It may fail to give off the signal when excessive radiation is present.
2. It may give off the warning signal (falsely) when excessive radiation is *not* present. This results in plant shutdown and careful checking of the reactor for the source of the (nonexistent) problem; the resulting loss of production and total expenses are quite large.

How can the designers and managers of the nuclear reactor cope with this two-faced monster? Keep in mind that money is no object, since malfunctions 1 and 2 are so serious and expensive themselves. A simple way

Table 4.1. Outcomes and Consequences for a Majority Rule System

Outcome	Action
None of the devices claim excessive radiation	No warning signal is given
Exactly one device claims excessive radiation	No warning signal is given
Exactly two devices claim excessive radiation, the third does not	A warning signal is given
All three devices claim excessive radiation	A warning signal is given

of reducing both dangers simultaneously is to install a complex of *three* such warning devices. The three devices are then linked so as to achieve a *majority rule*. Specifically, a control device is positioned to determine the output (or lack of it) of each warning system. Basically, it permits a warning signal to be emitted, when *two or more* warning devices claim excessive radiation is present. On the other hand if, for example, only one warning device claims excessive radiation is present, while the other two claim *it is not* (that is, disagree), then the control device abides by the majority rule—no warning signal is given. The term *majority rule* is certainly an appropriate descriptor for such a control device. To be perfectly specific, we list the possible outcomes and the resulting actions in Table 4.1.

Of course, whenever there is disagreement among the three detection devices, all three devices are very carefully examined to determine which device is malfunctioning. Appropriate repairs are made, not only on the minority opinion device, but if necessary on any (or both) of the other two devices. In addition, regularly scheduled checks or repairs are made on each of the warning devices and on the control device (the "poll taker"). This is to preclude the possibility that all three agree, but on the wrong decision, or alternatively, the poll taker itself is operating improperly.

How can we demonstrate quantitatively that the majority rule configuration (also called a *two-out-of-three* system) is superior to the single warning device, assuming decent reliability for the individual subsystems (that is, the three warning devices and the poll taker)? In the process of demonstrating superiority, we will also determine the *degree* of superiority as a function of individual device reliability.

By the **reliability** of a device we simply mean the chance that it will perform properly. In the present case where the warning device really has two functions, (a) issue a warning when necessary and (b) refrain from issuing a false alarm, for simplicity we will assume the device is equally reliable in performing the two functions. (Our aim, of course, is to demonstrate the statistical ideas in the simplest way, and not to confuse the reader with detailed complexities that obscure the basic principles.) We assume also that the control device is perfectly reliable, that is, it always correctly reports and reacts to the majority opinions among warning devices.

Suppose, then, that the chance that a single warning device reacts properly (gives a signal in response to excessive radiation and refrains from giving a false signal) is .99. What is the corresponding reliability if a two-out-of-three monitor is used? Suppose excessive radiation is actually present. The chance that all three warning devices detect and report the radiation is $.99 \times .99 \times .99 = .9703$. The chance that warning devices 1 and 2 detect and report the radiation but device 3 does not is $.99 \times .99 \times .01 = .0098$. A similar chance of .0098 holds for devices 1 and 3 detecting, while device 2 does not, and the chance is also .0098 that 2 and 3 detect, while device 1 does not. Since we have listed all four possible **mutually exclusive** outcomes leading to a warning signal (the correct reaction in this situation), we sum the four individual probabilities corresponding to a warning signal to get the total probability of .9997 $(= .9703 + 3 \times .0098)$ of a warning signal when excessive radiation is present.

Clearly, an exactly dual calculation can be performed when excessive radiation is *not* present. Note that the reliability .9997 of the two-out-of-three system is a considerably higher value than the .99 reliability achieved using a single warning device. Putting it another way, there is a reduction of *inappropriate* responses from 100 per 10,000 trials to only 3 per 10,000 trials.

Other Situations in Which a Majority Rule Device Is of Benefit

The principle illustrated previously, that *a majority rule device improves reliability*, is used in a wide variety of actually occurring situations. If the consequences of a lack of warning of an existent real danger are particularly disastrous, or if a false warning that a serious disaster is about to occur would lead to a very costly or destructive reaction, it may be well worthwhile to use a majority rule monitor of more than three warning devices; thus, it may be safer even though costlier to use a three-out-of-five system. Another factor that enters into the selection of the type of majority rule device that is most appropriate is a comparison of the chance of a false warning and its attendant cost with the chance of a lack of warning of an existent danger and its attendant cost.

Real-Life Examples

1. A warning device to alert the United States that a massive strike by enemy nuclear missiles is underway. Obviously, the consequences of either type of mistake are immeasurably costly.

2. An automatic landing gear for large commercial aircraft.
3. A device to warn of an imminent earthquake.
4. A device to warn of an impending hurricane.
5. A radar device to warn a train engineer or aircraft pilot of obstacles in his or her[†] path.
6. A heat- or smoke-sensitive alarm to warn residents of the breakout of a home fire, especially during the night hours. It is quite obvious in this case that the consequences of a lack of alarm when fire is present are far more serious than the consequences of a false alarm when no fire actually exists. This asymmetry is reflected in the probabilities of the two types of error resulting from the design of the system.

Many other examples exist. However, we have omitted from our discussion many of the complexities present in even the simplest cases.

Summary

1. The reliability of a device is the chance that the device will perform properly.
2. In warning devices designed to give off a signal when excessive radiation is present during operating of a nuclear reactor (and in many other safety devices), there are two types of errors, namely the "false-negative" error where the device fails to give off a signal when excessive radiation is present, and the "false-positive" error where the device gives off a signal when excessive radiation is not present.
3. A majority rule device, such as the one illustrated in the text, increases reliability.

Problems

1. An airplane functions if at least two of its three engines function. Each engine has a reliability (chance of functioning) of .99. What is the reliability of the airplane? (This is a simpler problem than discussed in the text, since only one type of failure is considered.)
2. For the nuclear reactor monitor, four detection devices are linked together. The control device gives off a signal if at least two out of the four detection devices "claim" excessive radiation is present; if only one or fewer detect excessive radiation, the control device ignores the minority detection and gives off no signal. The probability of detecting

---

[†] How will we say it when robots guide planes and trains?

excessive radiation is .99, while the probability of a false alarm is .01 for each individual device. The control device (the "brain") is perfectly reliable. What is the probability that the monitor
   a. Fails to detect excessive radiation?
   b. Emits a *false* warning (a warning when actually excessive radiation is not present)?
3. In Problem 2, plot the probability of a false alarm for the following values of the probability of a *false* detection of excessive radiation: .10, .05, .02, .001.
4. Suppose the control device is not perfectly reliable, but instead has a probability of .999 of reaching the right conclusion on each occasion when a decision must be made. In the text calculation, by how much do the errors of each of the two types increase?
5. Suppose that on the average, the nuclear reactor emits excessive radiation 10 times per year. With the monitor operating as described in the text, what is the average number of incidents per 10-year period in which excessive radiation occurs and yet no warning signal is emitted?
6.* Obviously, the detection and control devices must be checked and maintained regularly to keep them functioning properly. What is a reasonable maintenance policy to minimize both the frequency of undetected incidents of excessive radiation and of false alarms?
7. Describe other practical situations in which the two types of error are present. Does the game of poker lend itself to this type of analysis? In this case, the two types of error are deliberately induced.
8. The chance that a simple warning device reacts properly is .99, and the control device is perfectly reliable. What is the reliability if a three-out-of five monitor is used?
9.* Using additional warning devices increases safety but also increases cost. How would you approach balancing these two factors?
10. A manufacturer of warning devices asserts that a specific warning device reacts properly with a probability of .95. How can you check this claim?

Vignette 5

Should You Inspect Each Item Before Accepting the Lot?

For many years, inspectors tested or inspected each unit in a lot of product, before accepting or rejecting the lot. If relatively few defectives were discovered, the lot was accepted; if too many defectives were found, the lot was rejected. (An alternative procedure in the latter case was to remove the defectives found and accept the remainder of the lot.)

This procedure seemed reasonable and businesslike. You were not buying a pig in a poke. Rather, you knew exactly what you were getting when you purchased a lot, and perhaps more importantly, you were unwilling to purchase a lot which was not up to your standard of quality. Then along came Hitler, World War II, and an insatiable demand for endless supplies of product of all kinds. The luxury of 100% inspection could no longer be afforded—the former inspectors were now mostly in uniform *using* the product rather than *inspecting* it. The combination of inadequate person-power[†] for inspection, vastly increased need for product of acceptable quality, and severe constraints on money and time forced a complete change in the approach adopted toward product inspection and acceptance.

Fortunately, statistical procedures had already been developed by Dodge and Romig of the Bell Telephone Laboratories for *inspection of only a sample of items randomly selected from a lot as a basis for deciding whether to accept or reject the entire lot; moreover, the risks of accepting a "bad" lot or rejecting a "good" lot could be specified in advance and maintained.* See Dodge (1928, 1935) and Dodge and Romig (1929, 1941). The basic idea of the **statistical sampling acceptance plan** is simple: Suppose the lot contains 4,000 units of product. Suppose further that management decides that a lot is to be considered acceptable if 4% or fewer of the units in the lot are defective. A sample of 150 units is selected at

---

[†]The feminist objection to male-chauvinist language is justified. However, the alternative choices of words sometimes personhandle the writing.

random and carefully examined. If 11 or fewer defectives are found among the sample, the entire lot is accepted; if 12 or more defectives are found, the entire lot is rejected. (In statistical language, the **acceptance number** is 11 and the **rejection number** is 12.)

Note several interesting features of the sampling plan:

1. Only 150 units among the 4,000 present in the lot are actually inspected; this amounts to a *savings of over 96% of the inspection work* as compared with the traditional 100% inspection plan.

2. The plan realistically recognizes that a certain percentage (up to 4%, in the present case) of units may be defective and still the lot as a whole is considered acceptable.

3. By statistical theory, it is possible to determine (within a small error) the percentage of defectives in the entire lot of 4,000 by actually inspecting only 150 randomly selected units. This feature is responsible for the enormous savings in inspection labor.

4. By statistical theory, it is possible to calculate the chance of accepting a lot containing *any* specified percentage defective using the 150-unit sample plan. Thus Fig. 5.1 shows in the form of a continuous curve the probability of acceptance as a function of the proportion of the units in the lot that are defective. For example, the chance of accepting a lot containing 4% defectives is .98, the chance of accepting a lot containing 10% defectives is .20, the chance of accepting a lot containing 14% defectives is practically 0, and the chance of accepting a lot containing 3% defectives is practically 1 (corresponding to certainty). The curve plotted in Fig. 5.1 relating the probability of accepting the lot as a function of the

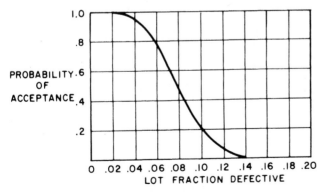

Figure 5.1 Probability of accepting a lot as a function of fraction defective (sample size 150, acceptance number 11, rejection number 12.).

percent defective in the lot is called the **operating characteristic** (OC) curve of the sampling plan.

Although statistical sampling plans had experienced quite slow adoption by American industry since their initial advocacy by Dodge and Romig in 1928, the pressures of World War II resulted in a rapid dissemination of knowledge concerning these plans and their widespread adoption throughout industry, the military establishment, and other branches of government. In fact, their use became worldwide.

There are sampling inspection plans that predate those of Dodge and Romig. Stigler (1977) gives a fascinating account of a sampling inspection scheme used to control the quality of the gold and silver coinage of the Royal Mint in London. The scheme, called the trial of the Pyx, has been in operation for about eight centuries. Stigler also points out that trials similar to that of the Pyx have been carried out by the U.S. Mint since 1792, the year the U.S. Mint was established.

The U.S. Department of Defense played a key role in the relatively rapid adoption of statistical sampling plans by industry in the United States and in other industrial nations. The department sponsored a series of short intensive courses in which the new methodology was explained to management, quality controllers, statisticians, and anyone concerned with product acceptance. In addition, the department developed and adopted as official procedure the use of *Handbook Military Standard 105A*, a set of statistical sampling plans accompanied by a careful, detailed explanation of their use.

H105A, as it was affectionately called by its aficionados, was almost as widely read by quality assurance personnel as the Bible and Harold Robbins. Essentially, it consisted of the following features:

1. A set of sampling plans
2. A set of corresponding operating characteristic curves
3. Rules for selecting among **normal, tightened,** and **reduced inspection**
4. A detailed explanation of the use of the handbook

Item 3 requires some explanation. Suppose the manufacturer has demonstrated over a long period of time an excellent record of product quality. It would seem reasonable to assume even before inspection of the next lot that quality is likely to remain good and therefore the sample size to be used could be reduced. This would save money and time spent on inspection. The reduced amount of inspection would continue so long as the manufacturer maintained a high level of product quality. Of course, if product quality showed signs of deterioration (for example, if a lot were rejected for containing an excessive number of defectives), the purchaser would shift back to normal inspection.

In a similar fashion, if the manufacturer has demonstrated over a period of time a poor quality level, it would seem reasonable to **tighten inspection**. This could be accomplished by requiring a larger sample size and/or a smaller percentage of defectives in the sample before permitting acceptance of the lot. If the manufacturer now demonstrates a consistent improvement in product quality, inspection would then revert from the tightened level to the normal level.

We have sketched some essential features of *one* important class of statistical sampling plans for acceptance of lots. A number of additional aspects of the subject are touched upon in the remarks that follow.

Remark 1. We have described briefly acceptance sampling plans in which the decision to accept or reject the lot is based upon the number of defectives observed in a *single* sample. More sophisticated plans are also available in which the inspector initially takes a first sample and counts the number of defectives present. If the number is relatively large, he rejects the lot; if relatively small, he accepts the lot; finally, if somewhere in between, he draws a *second* random sample from the lot. After inspecting the second sample, depending on the *total* number of defectives found in the two samples, the inspector either accepts or rejects the lot. This type of plan is called a **double-sampling** plan. On the average, it requires less inspection to reach a decision with the same risks as does a **single-sample** plan. The risks are: (1) the chance of accepting a poor lot and (2) the chance of rejecting a good lot. Either risk is, of course, the consequence of relying upon a sample rather than upon 100% inspection.

Military Standard 105D contains a set of double-sampling plans, along with instructions for their use. [Military Standard 105A has been revised several times; its current designation is Military Standard 105D. It is presented in full in Duncan (1974), along with a thorough discussion of its features.]

Remark 2. The logical extension of a double-sampling plan is a **multiple-sampling** plan. The inspector takes successive samples up to some number of samples *specified in advance*. Depending on the total number of defectives observed in the samples thus far taken, the inspector (1) rejects the lot, (2) accepts the lot, or (3) takes an additional sample. This **sequential procedure** continues (if necessary) until the maximum number of samples permitted has been taken. After inspecting the final sample, the inspector must accept or reject the lot; that is, he is not permitted to take an additional sample.

"Boss, only one defective in the entire lot!"

As you might conjecture, the *average* amount of inspection required is reduced even further as compared with that of the double-sampling plan. However, it is possible, for a few individual lots, to end up inspecting more units than would be required under the double-sampling plan.

It is also obvious that the administrative detail, chance of making mistakes, and level of inspector intelligence, diligence, and alertness required are greater under the multiple-sampling plan than under the double-sampling and single-sampling plans. This fact must be balanced against the reduction in the average amount of sampling. When inspection is quite costly (say, in testing or inspecting large or complicated systems), it may prove worthwhile to use multiple sampling. On the other hand, in inspecting nuts and bolts, the single-sampling plan may be chosen for its simplicity.

Military Standard 105D contains a set of multiple sampling plans.

Remark 3. Thus far our discussion has been confined to inspection in which the item is classified as defective or acceptable. For many products, a *measurement* is made, rather than a simple classification into "good" or "bad." For example, the resistance of a resistor might be measured. Statistical sampling plans have been developed for acceptance sampling of lots based on a measurement made on each unit in a random sample of product. Such plans, called **variables sampling plans,** are designed with essentially the same purposes in mind as the sampling plans discussed previously in which items are classified as defective or nondefective. These earlier plans are called **attribute sampling plans.**

When a critical variable is measured, rather than simply classified as either good or bad, it is intuitively clear that more information is being obtained. Thus, variables sampling plans tend to require smaller sampling sizes than attribute sampling plans. To offset this advantage, there is, of course, more sophistication required in both setting up the sampling plan and administering it on a routine basis.

In summary, the use of statistical sampling plans over the last four decades constitutes one of the major successes in the many applications of statistical methods. Many millions of dollars and thousands of personhours

VIRGINIA SMOLIAR

"General, our company guarantees absolutely every component. We do 100% destructive life testing."

have been and are still being saved. In addition, a large number of inspectors, quality engineers, and managers have become familiar with the methods and applications of statistics in industrial problems. Worldwide, almost every major industrial country has a quality control professional society; the total number of members is in the tens of thousands (the United States alone has roughly 20,000 members).

Summary

1. Using statistical sampling acceptance plans, inspectors can forego 100% inspection and still ensure acceptable quality.
2. The sampling plans maintain quality by controlling the risk of accepting a bad lot and the risk of rejecting a good lot.
3. The curve which relates the probability of accepting the lot to the percent defective in the lot is called the operating characteristic (OC) curve of the sampling plan.

Problems

1. Suppose inspection or testing the product manufactured results in its destruction (for example, inspection of bullets, mortars, rockets, or other explosive devices in the attribute sampling case; or estimation of life length of rubber tires or batteries; or breaking strength of fibers in the variables sampling case). Discuss the alternative inspection plans: 100% inspection versus statistical sampling plans. Hint: The answer is completely obvious.
2. From Fig. 5.1 determine:
   a. The chance of accepting a lot in which 5% of the units are defective
   b. The percent defective for which the chance of accepting the lot is 90 out of 100
   c. The percent defective for which the chance of rejecting the lot is 1/2
3. In the sampling plan described in this vignette, the sample size is 150. Suppose the consumer wishes to be more cautious—he now routinely inspects samples of size *300* and determines the corresponding acceptance and rejection numbers of defectives to achieve a probability of .98 of accepting a lot having 4% defectives (as in the original sampling plan). How do you think the new operating characteristic curve will compare with that of Fig. 5.1? More specifically,

    a. If the percent defective is relatively *large* (say 10%) which plan will reject the lot with higher probability, the original 150 sample size plan or the more conservative 300 sample size plan?

    b. If the percent defective is relatively *small*, which plan will *accept* the lot with higher probability?

4. (Problem 3 continued) Suppose the unit inspected is a battery, for which the quality characteristic of interest (say, the charge) can actually be *measured*, rather than merely classified as defective or not. Under an appropriate variables sampling plan, how would the sample size required to achieve the same assurance of accepting "good" lots and rejecting "bad" lots compare with the sample size of 150 originally selected?

5. Suppose you had to devise an attribute sampling acceptance plan for extremely expensive devices. Recalling that, compared with single-sampling plans, double-sampling plans reduce the average amount of inspection required, and that multiple-sampling plans reduce the average amount even further, suggest a type of sampling plan that would yield the maximum possible reduction in inspection.

6. It has been found from experience that 100% inspection of large lots may actually be *less* effective than inspection of a random sample (of appropriate size). What would explain this seeming anomaly?

7. Compare the two types of errors inherent with sampling acceptance plans with the two types of errors inherent with majority rule safety devices (as discussed in Vignette 4).

8. Suppose a manufacturer resubmits (unchanged) a lot that has already been rejected. What do you think of this practice? What, if anything, can the statistician or inspector do to prevent it?

9. Think of some instances in your life where you purchased a "lot" and did not use 100% inspection. Explain your formal (or informal) acceptance sampling plan.

10. What are some advantages of sequential procedures? What are some disadvantages of sequential procedures?

References

1. H. F. Dodge (1928). Using inspection data to control quality. *Manufacturing Industries 16*, 517–519, 613–615.

2. H. F. Dodge and H. G. Romig (1929). A method of sampling inspection. *Bell System Technical Journal 8*, 613–631.

3. H. F. Dodge (1935). Statistical aspects of sampling inspection plans. *Mechanical Engineering 57*, 645–646.

4. H. F. Dodge and H. G. Romig (1941). Single sampling and double sampling inspection tables. *Bell System Technical Journal 20*, 1–61.

5. A. J. Duncan (1974). *Quality Control and Industrial Statistics*, 4th ed. Richard D. Irwin Inc., Homewood, Ill.

6. S. M. Stigler (1977). Eight centuries of sampling inspection: The trial of the Pyx. *Journal of the American Statistical Association 72*, 493–500.

References 1–4 are cited here more for historical context than suggested reading. The reader may care to browse through reference 5 to get a glimpse of applications of statistics in industrial settings. Reference 6 is a fascinating account of the trial of the Pyx—the final stage of a sampling inspection scheme for the control of the quality of the product of the Royal Mint of Great Britain.

Vignette 6
Statistical Control Charts

During World War II, American industry made use of a "secret weapon" that saved millions and millions of dollars but, more importantly, significantly decreased the number of defective or substandard items of war material. These savings were accomplished without using additional personpower, money, or time. The secret weapon played an important part in winning the war.

The secret weapon was the **statistical control chart.** This simple graphical application of statistics developed to control the quality of material produced was actually introduced by Walter A. Shewhart of the Bell Telephone Laboratories as early as 1924 (see Shewhart, 1926a,b). However, its adoption by American industry proceeded relatively slowly until the shortage induced by war consumption and the ever-increasing demands of the military made it urgent to improve industrial productivity. During the war period (and subsequently) it was (and still is) used quite widely to ensure specified high quality of manufactured product. More recently, it has been used to control the quality of repetitive operations of a wide diversity, and not just confined to manufactured product (e.g., clerical operations, audit confirmations, and so forth)

What is a statistical control chart and how does it work? We illustrate the basic ideas with an example. A fragmentation bomb base is being manufactured by the American Stove Co. In order to meet measurement specifications set by the customer (the U.S. Army), the base height must lie between .820 and .840 in.; otherwise the assembled bomb may not function satisfactorily. To determine whether the production process is remaining under control throughout the day or, alternatively, is drifting out of control, random samples of five bomb bases each are selected at 15-minute intervals throughout the day, and the height is measured for each base. The detailed results are displayed in Table 6.1. In the last column, the average or mean height is recorded for each sample. For example, the mean of the first sample of five is computed as

Table 6.1 Overall Heights of Fragmentation Bomb Bases (Samples Taken at Random from Production of April 20, 1944; Specification, $.830 \pm .010$ in.)

Sample No.	A	B	C	D	E	Sample Mean
1	.831	.829	.836	.840	.826	.8324
2	.834	.826	.831	.831	.831	.8306
3	.836	.826	.831	.822	.816	.8262
4	.833	.831	.835	.831	.833	.8326
5	.830	.831	.831	.833	.820	.8290
6	.829	.828	.828	.832	.841	.8316
7	.835	.833	.829	.830	.841	.8336
8	.818	.838	.835	.834	.830	.8310
9	.841	.831	.831	.833	.832	.8336
10	.832	.828	.836	.832	.825	.8306
11	.831	.838	.844	.827	.826	.8332
12	.831	.826	.828	.832	.827	.8288
13	.838	.822	.835	.830	.830	.8310
14	.815	.832	.831	.831	.838	.8294
15	.831	.833	.831	.834	.832	.8322
16	.830	.819	.819	.844	.832	.8288
17	.826	.839	.842	.835	.830	.8344
18	.813	.833	.819	.834	.836	.8270
19	.832	.831	.825	.831	.850	.8338
20	.831	.838	.833	.831	.833	.8332
21	.823	.830	.832	.835	.835	.8310
22	.835	.829	.834	.826	.828	.8304
23	.833	.836	.831	.832	.832	.8328
24	.826	.835	.842	.832	.831	.8332
25	.833	.823	.816	.831	.838	.8282
26	.829	.830	.830	.833	.831	.8306
27	.850	.834	.827	.831	.835	.8354
28	.835	.846	.829	.833	.822	.8330
29	.831	.832	.834	.826	.833	.8312

Source: Lester A. Kauffman, *Statistical Quality Control at the St. Louis Divison of American Stove Company* (War Production Board, Office of Production, Research, and Development, *Quality Control Reports*, No. 3, August, 1945), p. 11.

$$\frac{.831 + .829 + .836 + .840 + .826}{5} = .8324.$$

In a similar fashion we compute the mean of each of the remaining 28 samples and list these means in succession in the last column.

The statistical control chart is shown in Fig. 6.1, displaying in succession each of the 29 sample means. Note that the middle horizontal

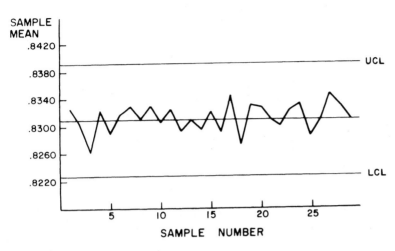

Figure 6.1   Control chart for sample means.

straight line is drawn at height (ordinate) .8312, which represents the average of the first 20 sample means. (The control chart was started at that point in the production process. Although 20 is a reasonable number of sample means to use in starting a control chart, it is certainly not a strict requirement. A somewhat smaller or larger number would also be acceptable.) In addition, the chart displays a pair of horizontal straight lines, one above the middle line and one below. The upper one is called the **upper control limit** (UCL), whereas the lower one is called the **lower control limit** (LCL). We will describe how these limits are computed shortly, but first let us see how the chart is used to control quality.

As successive means are plotted on the chart, we check whether any mean falls outside of the pair of control limits. If this should happen, then we take it as a warning signal that the production process is **out of control**. More specifically, if the current point plotted falls *above* the UCL, we conclude that the machine is tending to produce bomb bases of larger height than usual. At this point, production is halted, and the trouble-shooting engineer carefully examines the production process (machine settings), worker operations, materials being fed into the machine, and so forth, to pinpoint the cause(s) of the (unwanted) change in output quality. A similar sequence of steps is initiated if the current plotted sample mean falls *below* the LCL. Of course, the knowledge of which control limit has been violated helps the trouble-shooter in tracking down the cause of the out-of-control condition, especially if past experience is brought to bear upon the problem.

Another indication of production process trouble brewing is the display of a trend in the points plotted, even if all points are within control limits.

Rules vary among control chartists, but one simple rule is to look for trouble when a succession of, say, seven points in sequence are observed, each point above the preceding one (or below the preceding one). It can be shown that the probability of such a sequence (say, the rising sequence) is $1/(6 \times 5 \times 4 \times 3 \times 2 \times 1) = 1/720 = .0014$ if the process is still under control; a similar chance exists for the falling sequence if the process is still under control. Since the chance is small that a *controlled* process would give rise to such sequences, the chartist plays it safe and assumes that actually the process is *not* in control and calls in the trouble-shooter.

It is apparent that the control chart can be a very useful and simple device for controlling the quality of product in a timely fashion. That is, the alert chartist can prevent the machine from continuing to produce unacceptable material over any significant length of time by promptly calling in the trouble-shooter at the first clear-cut indication of an out-of-control state (that is, a point outside of control limits, or a succession of rising or falling points of specified length).

How are the upper and lower control limits (.8395 and .8229) calculated? The basic idea is to set the upper control limit at a value such that, if the machine is actually operating as it does on the average (that is, producing bomb bases having a height of .8312), then only a very small fraction of the time, purely by chance, a sample of five will be produced having a mean value greater than the UCL; similar reasoning governs the computation of the LCL. In computing the UCL and the LCL, the statistician bases his calculation upon a sufficient number of observed samples to determine fairly closely the proper values of the limits (in the present example, 20 samples are used). Fortunately, by using statistical theory, he does not require anything like 100 sample means, but merely 20-30, to arrive at fairly accurate UCL and LCL values.

The alert reader may quite validly raise the question, What is the rationale for setting the UCL (or alternatively, LCL)? Should the limits be influenced by economic factors, or in some cases, by safety factors (say, in monitoring a nuclear reactor)?

An excellent question! The answer is yes, the limits should be custom-made for each production process by balancing the costs of (1) setting off a false alarm (when, purely by chance, a point falls outside the limits even though the process is operating perfectly satisfactorily), and (2) *not* setting off an alarm even though the process has slipped out of control; this would happen if the points plotted remained between control limits purely by chance. By balancing these costs optimally to achieve a minimum total cost over the long haul, the control chartist might very well calculate control limits which differed from the standard ones, and which in fact, might not even be symmetric about the nominal values. In cases in which human safety is involved (say, the production of noxious gases, explosives, nuclear energy, or the monitoring of an airplane flight), the argument for the individualized calculation of control limits becomes even stronger.

However, most control charts plotting the successive sample means are designed in a fairly standard fashion. One of the main reasons is that many companies, governmental units, and other organizations using control charts do not have sufficient statistical expertise to custom-tailor individual charts. The control chart user is satisfied to follow the statistical handbook directions for setting up a "standard" control chart.

One of the key reasons for the great and widespread success of control charting is the simplicity of calculation, plotting, and maintenance of the chart. This permits any machine operator who can add the ability to maintain the control chart. In addition, the machine operator generally derives considerable satisfaction from the feeling of being in greater control of his machine and being continually aware of the state of its operation. A greater pride and an increased involvement in his work are engendered, thus leading to better quality output and fewer defective parts produced. Thus, in addition to the obvious benefits of a measurable statistical type, there also are benefits of a more subtle psychological type. Finally, the use of control charts throughout a factory serves as a simple, unified management tool; specifically, the quality control engineer who may have originally designed the statistical control charts will generally conduct a tour of the factory at regular intervals to obtain a direct, firsthand, *quantitative* knowledge of the status of product quality throughout his area of responsibility. During the tour he can also observe how well the charts are being maintained. The more experienced, statistically trained engineer will sometimes detect a chart that is too good to be true. That is, for one reason or another a particular chartist may decide that it is necessary or more convenient to invent the values of successive points for his chart than to go to the trouble of selecting five units at random from the production output, measuring each unit, adding the five measured values, dividing the sum by 5, carefully plotting the resulting sample mean on the chart, and then deciding whether the production process should be considered in or out of control. Instead, the con chartist invents a sequence of points which he believes will show that the chart is under control, but which also shows the usual statistical fluctuations due to chance.

Unfortunately, he may have insufficient statistical training to simulate a random series. (Strangely enough, it is not easy to invent or simulate a random series of numbers; apparently, the human mind imposes a regularity which may often be detected.) For example, the quality control engineer may run across a chart that appears as Fig. 6.2.

It is readily apparent to the engineer that there is something fishy about the chart. It displays far too much regularity to have evolved through chance

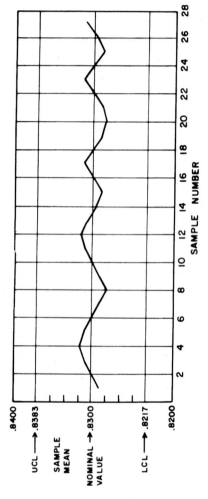

Figure 6.2   The chart of a con artist.

mechanisms. Note (1) the cyclic appearance of the pattern of successive points; (2) the rise or fall between successive points is about the same throughout the chart; (3) the relatively small **range** of the set of points (i.e., difference between the largest and smallest values in the set); (4) the large number (seven) of points that fall exactly on the nominal value (26% of the plotted points); and (5) that successive cycles are almost equal in length.

A short friendly discussion between the quality control engineer and the chartist reveals that the raw data somehow have been "lost" and that actually the points on the chart were invented. Further discussion reveals that the chartist had missed attending the educational sessions in which chart maintenance had been explained. The quality control engineer then spent some time with the inventive chartist carefully explaining the entire procedure in detail and observing the chartist carry it out successfully for a sufficient number of points. All ended well—the errant chartist dropped a load of guilt and picked up instead a pride in the newly developed control over machine output quality. The engineer derived satisfaction from his skillful handling of a ticklish situation.

We have not discussed the upper and lower **tolerance** values, .840 and .820, respectively, and their relationship, if any, to the upper and lower **control** limits. The tolerance values .840 and .820 represent the largest and smallest values of the bomb base height acceptable to the customer, whereas the control limits .8395 and .8229 represent the critical values which alert the machine tender to look for trouble. The tolerance limits are customer imposed, whereas the control limits describe machine capability. Of course if the control limits are further apart than the tolerance limits, we face a serious problem—*the customer is requiring quality which the machine is incapable of providing as presently operated.* Thus, the control chart provides crucial information as to the proportion of unacceptable units that will be produced even when the machine is operating under good control. If this proportion is too high, something must give—either the customer reexamines his tolerance limits and relaxes them; or the production process is carefully studied, and the current random variation is reduced by appropriate improvements in either machine operation, worker care, or raw material fed into the machine.

Summary

1. The statistical control chart is a useful and simple device for timely control of the quality of repetitive operations, such as a production process.

2. The control chart gives a graphical display of the sample means of the process.

3. The control chart contains a pair of horizontal lines, the upper control limit (UCL) and the lower control limit (LCL). If a sample mean falls outside the control limits (that is, above the UCL or below the LCL), this is to be interpreted as a warning that the process is out of control.
4. A trend in the sample means, often easily detectable from the chart, is another indication that the production process may have gone awry.

Problems

1. Suppose a certain chemical product is being manufactured. A control chart is to be kept on an important quality characteristic, say the purity of the chemical product (that is, absence of foreign materials). A nominal value for product purity has *not* been set by the customer, since the product is a consumer product and will be purchased by millions of individual customers. Assuming the production process has been carefully designed, worker operation is carefully controlled, high-purity raw materials are fed into the process, and all reasonable precautions have been taken to keep the entire operation free of impurities, how can the control chartist calculate a "central" value (in place of a preset nominal value) from the successive sample means being plotted?

2. A manufacturer of a precious metal wishes to ensure that the units he produces are as alike as possible in a certain critical aspect. Could he set up a control chart as follows? At regular intervals, the machine tender draws a sample of five units at random. He measures the critical quality characteristic for each unit. However, instead of computing the sample mean, he computes the **sample range,** that is, the difference between the largest and the smallest values in the sample. For example, if the five measured values in his sample were 2.83, 2.79, 2.81, 2.83, and 2.86, the sample range would be 2.86 (the largest) − 2.79 (the smallest) = .07. Now he proceeds to plot a control chart for the *sample range,* rather than for the sample means. Since he is concerned about *excessive variability* in product quality, he computes an upper control limit for the sample range. [For each specified sample size, an appropriate upper limit can be determined as a multiple of the average sample range (see Duncan, 1974)]. He also displays the **average sample range** based on the first 20 samples as a central horizontal line; this serves as an indication of the usual behavior of the sample range when the process is under control with respect to product variability. The chart appears as in Fig. 6.3.
   a. Does the process remain in control with respect to product

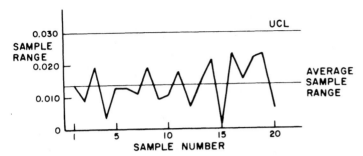

Figure 6.3   A sample range chart for controlling product variability.

variability throughout the period in question? If not, which samples indicate excessive product variability?
b. What could explain excessive product variability?
c. Two types of error may occur in the course of using a control chart for the sample range. Describe each.
d. Several alternative measures of sample variability are available, including the well-known and commonly used **sample standard deviation** (which we won't define here; either you know it or you don't—for our present purpose, it does not matter). Why is the sample range generally used in control charting?

3. Suppose in the bomb base manufacturing process discussed in the text, the calculated UCL was considerably higher than the upper tolerance limit (.840), and the calculated LCL was considerably lower than the lower tolerance limit (.820).
    a. Would you expect the proportion of out-of-tolerance product to be large?
    b. Would it serve any purpose to stop the production process and trouble-shoot each time a sample point fell outside of tolerance limits (not *control* limits)?
    c. What conclusions would you draw from a comparison of control limits and tolerance limits?

4. The additional observations shown in Table 6.2 were taken on bomb base height in the next 10 samples in the production process charted in Fig. 6.1.
    a. Compute the mean of each sample.
    b. Plot the sample means in succession, thereby continuing the control chart in Fig. 6.1.
    c. Which sample means indicate out-of-control conditions?

Table 6.2   Bomb Base Heights of Additional Samples

Sample Number	Observation A	B	C	D	E	Sample Mean	Comment
30	.832	.836	.829	.830	.832		
31	.835	.836	.839	.832	.837		
32	.838	.842	.845	.843	.845		Process adjusted
33	.831	.829	.832	.828	.827		
34	.826	.827	.831	.829	.830		
35	.828	.825	.832	.837	.825		
36	.832	.827	.828	.830	.833		
37	.829	.831	.834	.828	.827		
38	.826	.820	.823	.827	.822		
39	.818	.810	.812	.813	.820		Process adjusted

5. Can you devise some other reasonable measures of sample variability besides the sample range? Describe these measures.

6. How would you create a control chart to ensure the quality of tennis balls continuously produced?

7. Describe some situations in which control charts may provide a clue as to what specific defects are present in the machine producing the product.

8. What problems could arise if a quality control engineer decided to maintain control charts only every other day of the production process?

9. What conclusions might you draw from a control chart that first showed an increasing trend and later showed a decreasing trend?

10. How can statistical control charts and sampling acceptance plans (Vignette 5) be coordinated to improve efficiency and quality, thus benefiting both the manufacturer and the consumer?

References

1. A. J. Duncan (1974). *Quality Control and Statistics*, 4th ed. Richard D. Irwin, Inc., Homewood, Ill.

2. W. A. Shewhart (1926a). Finding causes of quality variations. *Manufacturing Industries 11*, 125–128.

3. W. A. Shewhart (1926b). Quality control charts. *Bell System Technical Journal 5*, 593–603.

Reference 1 is worth examining to get an idea of the scope of industrial applications of statistical methods. References 2 and 3 are included here more for historical context, rather than suggested reading.

Vignette 7
If One Is Good, Why Not Have Two? Redundancy

For want of a nail, the shoe was lost,
For want of a shoe, the horse was lost,
For want of a horse, the rider was lost,
For want of a rider, the battle was lost,
For want of a battle, the kingdom was lost,
And all for the want of a horseshoenail.

Margaret de Angelis, *Mother Goose*

They also serve who only stand and wait.

John Milton, *Sonnet XIX: On His Blindness*

The years, however, had taught him caution; an exchange was like the workings of a giant aircraft. Each system has a backup system, each backup an alternative.

Robert Ludlum, *The Matarese Circle*

She was sporting a round badge saying "Save the Whale" and that was the other sign, said Tayeh, because from now on Khalil requires that there always be two things: two plans, two signs, in everything a second system in case the first fails; a second bullet in case the world is still alive.

John le Carré, *The Little Drummer Girl*

How many lives have been saved because humans have *two* kidneys and *two* lungs? How many people can still see even though blind in one eye, and can still hear even though deaf in one ear? How many airplanes have made a safe landing even though one or more engines failed or were shut down to prevent fire?

51

The point is clear: In both biological systems and man-made systems, for added protection it may be quite prudent and advantageous to design the system so that even though one subsystem may suffice to perform a certain function, two (or even more) copies may be provided. Thus, the extra subsystem is available to continue the function (usually a vital one for system survival), even though one of the subsystems fails. We call the extra subsystem **redundant**; that is, its main purpose is to ensure that the function in question continues. True, it may be helpful in other ways; for example, the second engine in the airplane may provide greater power, speed, and control, the second eye in the human may give the ability to perceive in depth, and so on. (Note that even in communication, a certain amount of redundancy is helpful—in the preceding sentence, the airplane example would have been enough to bring out the point, but the addition of the human eye example ensures that the point is not missed by a hasty reader.)

Redundancy may be incorporated into a system in one or more ways. In the examples listed previously, the extra subsystem operates *simultaneously* with its partner. In other cases, redundant subsystems or components may simply stand by, that is, serve as spare parts to be used to replace failed components. For example, the spare tire generally present in an automobile is called into use upon failure of any of the four operating tires, but at other times simply is kept available for the replacement of a failed tire. Thus, it experiences essentially no deterioration during those other times.

To distinguish between the two types of redundancy, we shall call the first type **active redundancy** (in which both the original component and the additional component are actively functioning), and the second type **standby redundancy** (in which only the original component is functioning, while the additional component is standing by and not in active use).

How much do we gain by utilizing active redundancy? That is, how much of an increase in reliability results from providing an additional active component?

Let us consider a simple illustration. Suppose a *single* radar has a chance of .99 of functioning. Since the target being guarded by the radar is of vital importance, the commanding officer wishes to increase the chance of detecting an enemy missile by installing an *additional* actively operating radar. What will the improved chance of detection become if he should carry out his plan? [To keep the example simple, we assume that an enemy missile, if present, is detected with perfect assurance by a functioning radar, that failure of a radar is recognized as soon as it occurs, and that a failed radar is switched off, with the second (redundant) radar continuing to operate.]

The calculation is most simply performed by first noting that the chance that the first radar fails is $1 - .99 = .01$; a similar chance exists for the second radar, if present. Thus, the chance that both radars fail is $.01 \times .01 = .0001$.

Finally, the chance that at least one of the radars *does not* fail is the complementary probability, $1 - .0001 = .9999$.

Thus, by installing the second radar, the commanding officer increases the chance of detection from .99 to .9999; perhaps a more dramatic version of the same conclusion is that the chance of an enemy missile entering undetected drops from 1 in a 100 to 1 in 10,000 by the addition of the second (actively redundant) radar.

Suppose now that the commanding officer were to plan to use the second radar, not in the actively redundant fashion, but in the standby fashion. That is, the second radar would be standing by ready to be switched on if the first radar failed. Clearly, this standby redundancy would differ in several respects from the active redundancy considered previously—some of the differences being disadvantageous, and at least one of them being advantageous.

Disadvantages of using the second radar in the standby fashion are:

1. In actual practice, some time would be required during which the failure of the first radar would be detected and the initiation of active operation of the standby radar. During this "dead time," an enemy missile could enter undetected.

2. The standby radar could not remain completely quiescent throughout the period preceding its call to active duty. Regularly scheduled tests would need to be performed to ensure that the standby radar remained in satisfactory operating condition.

There is at least one great advantage in using the second radar in the standby redundancy fashion: The sum of the lifelengths of the two radars operated in the standby redundancy mode will be greater than the longer of the lifelengths of the two radars operated in the active redundancy mode. This is obvious since in the active redundancy mode the two radars are simultaneously operating, while in the standby redundancy mode first one radar is operating, and then upon its failure, the second radar takes over.

The term *standby redundancy* is more readily recognized by most people under the more familiar name *spare parts provisioning*. In this context, we recognize the important problem faced by many decision makers of determining the number of spare parts that should be provided for each of the critical components of a complex system, especially a system not easily resupplied. For example, a space vehicle will be making a trip of some duration with three people aboard. It is to be expected that failures or wearout of some critical parts will occur. From previous experience with the critical components, the engineers have a reasonably good idea of the statistical lifelength of each critical component, and know almost exactly the weight of each. The basic problem of the decision maker is to determine the number of spares of each critical component type to provide in the space vehicle so as to assure with very high probability that no shortage occurs during the space trip and yet keep the total weight of the vehicle plus spares under a critical total weight.

Such problems are solvable with the information specified. The solutions have been used in a variety of applications, such as (1) provisioning spare parts for airplanes at airports, (2) determining optimal spares kits for military systems that are to be tested in the field, (3) outfitting nuclear submarines with spares for a long mission before returning to port, and so forth (see Barlow and Proschan, 1965, chap. 6).

Selecting Type of Redundancy Design

Design engineers often face a choice in selecting the type of redundancy which is best for a given system or subsystem. We illustrate with a simple example which brings out the principle involved.

Example. An amplifier circuit must be designed to achieve a gain of 100. The "minimum" circuit which will achieve this gain is shown in (a) of Fig. 7.1. It consists of two small amplifiers (1 and 2); each small amplifier magnifies the volume of sound by a factor of 10. Two alternative circuits are shown in (b) and (c) which will achieve the desired gain, but with redundancy of different types. Suppose an individual small amplifier has a **reliability** (chance of functioning properly) of .9. Let us compute the reliability of each of the three circuits, and compare them.

The minimum circuit displayed in Fig. 7.1(a) has a reliability of $.9 \times .9 = .81$, since each of the two small amplifiers, 1 and 2, must function to achieve a combined gain of 100 for the circuit.

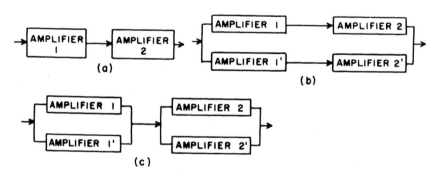

Figure 7.1 Minimum circuit and two redundant circuits for amplifier.

VIRGINIA SMOLIAR

"This tape describes the advantages of using redundancy, click, redundancy, click, redundancy, click, . . . "

The redundant circuit in (b) has a reliability of $1 - (.19)^2 = .9639$. To see this, note that in order for the circuit to fail, both subsystems (1,2) and $(1',2')$ must fail. The chance of both subsystems failing is $(.19)^2$. Finally, the reliability of the circuit is simply the complementary probability $1 - (.19)^2$. Thus, the reliability of circuit (b) is greater than that of circuit (a) because of the redundancy built into (b).

Circuit (c) is the most complicated in appearance; is it the most reliable? It is easy to verify that indeed circuit (c) is the most reliable. Simply comparing circuit (c) with circuit (b), we see that if, for example, components 1, 2, and 1' function in (b), then the circuit in (b) functions. If the same trio of components in (c) functions, then the circuit in (c) also functions. More generally, any time (b) functions, (c) also functions. On the other hand, the failure of components 1 and 2' ensures the failure of circuit (b), but *not* of circuit (c). Thus, circuit (c) must be more reliable than circuit (b).

It is clear that the reliability of a system does not depend only on the amount of redundancy, but also on how the redundancy is allocated throughout the system. One general principle used by design engineers may be stated as follows. *All other things being equal, redundancy at the component level yields greater reliability than redundancy at the system level.*

This principle holds for a great variety of systems, both engineering and biological. It can be proved mathematically. It is important to note, however, that in some applications all other things are *not* equal. For example, it may

"My son, are the Westerners discovering the advantages of redundancy just now?"

require many more switches to achieve redundancy at the component level than at the system level. These additional switches may introduce unreliability because of their possible failure. On the other hand, if the switches have much higher reliability than do the components, then the principle is quite useful.

A simple example of the principle occurs in Fig. 7.1. Note that in (b) redundancy of the *system* of two amplifiers (1 and 2) is displayed, whereas in (c) redundancy of each *individual* amplifier is displayed. We have already seen that system (c) is more reliable than is system (b).

Summary

1. Extra subsystems (or components), called redundant subsystems (components), provide added protection in both biological systems and man-made systems. These extra subsystems are present to ensure that the desired function in question continues.

2. Active redundancy refers to the situation in which the extra subsystem operates simultaneously with its partner. Standby redundancy refers to

the situation in which the original subsystem is functioning and the additional subsystem is standing by and not in active use.

3. Redundancy is used to increase reliability.
4. The reliability of a system depends not only on the amount of redundancy, but also on how the redundancy is distributed throughout the system.

Problems

1. A nuclear reactor is being monitored for radiation emission by a single detection device. The chance that it fails to detect radiation emission is .01. After the nationally publicized Three Mile reactor accident, the manager installs an additional detection device of the same type to operate simultaneously with the original device. What is the chance now that hazardous radiation emission will go undetected because of failure of both devices? (Recall Vignette 4, "Monitoring a Nuclear Reactor.")

2. The situation described in Problem 1 has another important aspect. The radiation detection device also has a chance of .005 of sending an incorrect signal that radiation is present when actually it is not. What is the chance that
   a. Both detection devices fail in this mode?
   b. Either one sends a false warning signal, while the other performs properly (that is, does not issue a false warning signal)?

   (Note that the chance of receiving a false warning signal is actually *increased* by the installation of the redundant detection device.)
   Can you suggest any alternatives that might reduce simultaneously the chance of experiencing either type of error (that is, not detecting radiation when it is present, or believing radiation is present when actually it is not)? Keep in mind that the devices are not expensive when compared with the cost of undetected radiation or the cost of shutting down a nuclear reactor unnecessarily, evacuating the premises, and looking for the cause of a nonexistent radiation leak.

3. An amplifier circuit is designed to achieve a gain of 1,000. It consists of three small amplifiers as shown in the primitive circuit in Fig. 7.2a; each small amplifier magnifies the volume of sound by a factor of 10. The reliability of each small amplifier is .95. What is the reliability of the amplifier circuit?
   Suppose another copy of the amplifier circuit is designed to serve in active redundancy with the first, as shown in Fig. 7.2b. What is the reliability of the redundant system shown in (b)?

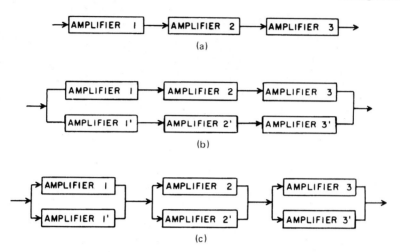

Figure 7.2 Amplifier systems.

Finally, suppose active redundancy is introduced at the component level as shown in Fig. 7.2c. What is the reliability of the system in (c)?

Comparing reliabilities of the three systems, what conclusions do you reach?

4. Describe some examples of redundancy in
   a. Engineering systems
   b. Biological systems
   c. Social organizations
   d. Verbal and written communication

   Can you find any examples of redundancy in this vignette on redundancy?

5. In this vignette, we have essentially described the value and advantage of using redundancy. Is it possible to incorporate *too much* redundancy into a system? Give examples of such excesses in a variety of contexts.

6. Describe some examples of redundancy in the world of sports.

7. Describe a type of redundancy (let's call it intermediate redundancy) which is intermediate to active redundancy and standby redundancy.

8. In most of our examples, the system is either functioning properly or it is not functioning properly (we could call this a two-state system). Describe some situations where one is dealing with a multistate system.

9.* Refer to Problem 8. What are some possible definitions of reliability when dealing with a multistate system?

10. Describe some situations of active redundancy in which:

a. Stress on the original component does not affect the performance of the additional component(s)

b. Stress on the original component does affect the performance of the additional component(s)

11. Redundancy is often used in communication to stress a point being made. Thus, although at this time it is grammatically incorrect, some less educated people will make a statement of the type: "I certainly will not take no cut in pay!" The use of the double negative has become so common in speech that it may very well become acceptable.†

References

1. R. E. Barlow and F. Proschan (1965). *Mathematical Theory of Reliability.* Wiley, New York.
2. R. E. Barlow and F. Proschan (1975). *Statistical Theory of Reliability and Life Testing: Probability Models.* Wiley, New York.
3. M. L. Shooman (1968). *Probabilistic Reliability: An Engineering Approach.* McGraw-Hill, New York.

References 1–3 require advanced mathematics to follow the technical derivations. Yet many readers will benefit from browsing through them to get an idea of the nature of reliability problems and of their importance.‡

---

†F. P. has labeled this Problem 11. Yet no problem is stated. Problem: What is the problem with F. P.? M. H.

‡References 1 and 2 are the best books ever written on statistical applications. F. P.

F. P. sometimes (actually, always) gets carried away in his fervent reverence for his own books. Objectively, I happen to think *Statistics: A Biomedical Introduction* by B. W. Brown, Jr., and M. Hollander is a far better book on statistical applications. M. H.

Vignette 8
Predicting the Reliability of a Complex System

MIAMI —An Eastern Airlines L-1011 jet carrying 172 people between Miami and Nassau lost its engine oil, its power and 12,000 feet of altitude over the Atlantic Ocean on Thursday before one of its three engines could be restarted and a safe landing made.

Each engine was afflicted by a mysterious and nearly simultaneous loss of oil pressure and supply. Late in the afternoon, Eastern announced the discovery that an oil seal was missing from each engine. . . .

Traub said the crew was warned of low oil pressure and supply by indicator lights like those on the dashboard of a car. At first, they thought something was wrong with the lights.

"They considered the possibility of a malfunction in the indication system because it's such an unusual thing to see all three with low pressure indications," Traub said. "The odds are so great that you won't get three indications like this. The odds are way out of sight, so the first thing you'd suspect is a problem with the indication system."

From an article by Michael H. Cottman and Arnold Markowitz, *Miami Herald*, May 6, 1983

MIAMI (AP)—Mistakes by two mechanics caused all three engines of an Eastern Airlines jumbo jet to stall, dropping the aircraft perilously close to the Atlantic Ocean, federal investigators said Friday.

One mechanic failed to install six tiny rubber seals on the engines' oil plugs during routine maintenance on the L-1011, leaving a gap for all of the oil to leak out when the engines were fired up, National Transportation Safety Board spokesman Ira Furman said.

Tallahassee Democrat, May 7, 1983

MIAMI (AP)—The spokesman gave this account of the maintenance lapse:

In the back shop Wednesday night, a mechanic pulled out the magnetized oil plugs for a routine check done after 25 hours of flying. The magnet collects metal from the oil, and mechanics analyze the chips to see if any engine parts are wearing abnormally.

The supervisor was out of replacement plugs, so that mechanic went to the stock room to get one for each of the three Rolls Royce RB-211 engines.

He and a partner screwed the plugs in, neither of them noticing that the sealer rings, called "O rings," were missing. The omission apparently didn't show up during a brief oil-pressure test.

The mechanic might have forgotten, Furlow said, because plugs kept at the supervisor's desk are equipped with the rings. Stock room plugs are not.

Tallahassee Democrat, May 8, 1983

In the early stages of development of a complex system, a great deal of interest exists among system designers, project managers, and future system users in predicting the reliability of the final system. Knowledge of system reliability will be helpful in planning the number of systems required, in providing optimal numbers of spares, appropriate maintenance facilities, determination of contractual requirements, and so forth.

How do reliability analysts predict the reliability of a complex system from a knowledge of system structure and a knowledge or estimate of component reliabilities?

Let us consider an example. A missile is being designed. The major subsystems of the missile are the propulsion subsystem, the guidance subsystem, the communication subsystem, and the warhead. Each of these subsystems must function if the missile is to function. Thus, the reliability of the missile is the product of the following four factors:

1. The reliability of the propulsion subsystem
2. The reliability of the guidance subsystem
3. The reliability of the communication subsystem
4. The reliability of the warhead

The next task of the analyst is to draw diagrams showing how each of the four subsystems is constructed; that is, how the components in each subsystem are linked together to form the subsystem.

For example, suppose the communication subsystem consists of the following components linked together as shown in Fig. 8.1.

Note that the two receivers are used in **parallel**; that is, if either or both functions, the system functions. This serves two purposes.

1. If one of the receivers fails, the other can still serve to receive signals. This redundancy increases the reliability of the communication subsystem (see Vignette 7).

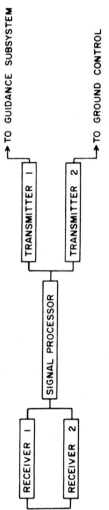

Figure 8.1 Assembly of components of communication subsystem.

2. The two receivers are oriented differently so that receiver 1 picks up signals from certain directions more clearly than does receiver 2, while receiver 2 picks up signals from other directions more clearly. The functioning of the pair of receivers may be compared with that of the two ears possessed by humans.

Carrying the analogy a bit further, we note that the signal processor serves a function similar to that of the brain: it receives the information, digests it, and transmits instructions to the guidance system via transmitter 1. Transmitter 2 sends information to the ground control station, such as the location of the missile, its velocity, its direction of motion, and so on.

Suppose, from a good deal of prior experience with similar components, the analyst knows the reliability of each of the components of the communication subsystem. Specifically, he knows that the reliability of each receiver is .95, of the signal processor is .98, and of each transmitter is .99. From the description of the functions of the components, we conclude that the communication subsystem will function properly if:

Event (1); at least one of the two receivers functions
and
Event (2): the signal processor functions
and
Event (3): both transmitters function.

First we compute the probability of event (1). The simplest way is, oddly enough, to first compute the probability $.05 \times .05 = .0025$, that both receivers *fail*. Subtracting .0025 from 1 yields .9975, the probability that at least one receiver functions. [Clearly, *either* (A) both receivers fail *or* (B) at least one receiver functions; it can't be true that events A and B both happen.]

The probability of event (3), both transmitters function, is $.99 \times .99 = .9801$.

Multiplying together the probabilities of events (1), (2), and (3) yields $.9975 \times .98 \times .9801 = .9581$, representing the probability that the communication subsystem functions.

In a similar fashion, we can compute the reliability of the propulsion subsystem, of the guidance subsystem, and of the warhead. Suppose, for example, these reliabilities turn out to be .97, .975, and .995. Then since the missile functions if, and only if, the propulsion subsystem, the guidance subsystem, the communication system, and the warhead all function, we compute the reliability of the missile as:

Missile reliability $= .97 \times .975 \times .9581 \times .995 = .9016$.

We may now summarize the method of computing system reliability for complex systems in general. First, determine the subsystems of the system. These subsystems and their relationship may be displayed in a system diagram: (Often the system is formed as a **series** arrangement of the subsystems: that is, the system functions if, and only if, each subsystem functions.) Next, represent diagrammatically each subsystem as an appropriate arrangement of the components comprising that subsystem. For each component, show diagrammatically how it is assembled from the parts comprising it.

At this step, usually the analyst is in a position to estimate or predict the reliability of each of the parts comprising any given component. For many standard parts, he may obtain part reliability from handbooks issued by manufacturers, government agencies (usually in the Department of Defense), or professional societies. For recently or newly developed parts, such information may not be available. In such cases, the reliability analyst may have to design a testing program to measure part reliability or, alternatively, draw upon past experience with similar parts and, in addition, make adjustments for the possible added complexity of the new part.

Using his estimates of part reliabilities, the analyst then computes component reliabilities from the component diagrams. Next, he uses these calculated component reliabilities to compute subsystem reliabilities from the subsystem diagrams. Finally, he uses the calculated subsystem reliabilities to compute the system reliability. If, as is often the case, the system is formed as a series arrangement of the subsystems, then system reliability is simply the product of subsystem reliabilities.

It is important to keep in mind that this **prediction** of system reliability is made in the early stages of system development. During the development of the system, tests are run on the components, subsystems, and the system as a whole, and these tests provide estimates of system reliability. These estimates of system reliability should be distinguished from the predictive values earlier calculated. The estimates obtained from the actual tests are more trustworthy in that they are based on fewer assumptions and more directly on observations of unit success or failure. At the end of system development, that is, just before the system is to be delivered to the purchaser or just before the system is to go into routine production, further tests are performed on the system as a whole. From these tests, the analyst is likely to obtain the most trustworthy estimates of system reliability. The tests on the system as a whole may reveal causes of failure due to the *interaction* among subsystems which were not predicted or visualized earlier, and which therefore were not incorporated in the reliability prediction.

One error that can lead to inflated estimates of reliability is the often unjustified assumption of *independence* between subsystems. The assumed independence may fail to be valid because the subsystems operate in a *common environment.* As an example, consider the L-1011 jet whose three engines stalled (see the news excerpts at the beginning of the vignette). The engines were subject to a common environment, one which allowed for the same O-ring error to be made on each engine during routine maintenance. This error caused *dependent* failures. Yet Traub said " . . . the odds are way out of sight . . . " because he was using an incorrect independence-based intuition.

Summary

1. The reliability of a complex system depends on the reliabilities of its subsystems and on how these subsystems are arranged.
2. Often the system is formed as a series arrangement (that is, the system functions if and only if each subsystem functions).
3. In a series arrangement of subsystems, the system reliability is obtained by multiplying the subsystem reliabilities.

Problems

1. A hi-fi system is composed of subsystems arranged as shown in Fig. 8.2. We wish to determine the probability that the system is capable of producing music, either stereophonic or monaural, either through records or through the radio tuner.

     The subsystems are standard and have been in use for some time. The reliability of each subsystem is known from past experience; the subsystem reliabilities are as follows:

Subsystem	Reliability
Tuner	.99
Changer	.98
Amplifier	.995
Speaker A	.98
Speaker B	.98

Figure 8.2   Arrangement of subsystems of hi-fi system.

Calculate the probability that the system is capable of producing music.

2. A computer system consists of a computer in series with two electrical power generators in parallel, as shown in Fig. 8.3.

The reliability of each of the three subsystems has been calculated by studying the arrangement of the components in each subsystem and making use of the known reliabilities of the components. The subsystem reliabilities are:

Subsystem	Reliability
Computer	.97
Generator 1	.99
Generator 2	.99

a. Determine the reliability of the system as a whole from a knowledge of the subsystem reliabilities just listed.

b. The **unreliability** of a device is the probability that the device fails.

Suppose by constructing the computer system of better (and, of course, more expensive) components, the unreliability of each subsystem is cut in half (for example, the unreliability of generator 1 is reduced from .01 to .005). What is the new, improved system reliability?

3. A single fire alarm system has a probability of .99 of emitting a warning when fire is present. To increase the probability of detecting a fire in a nuclear plant, the plant manager installs two fire alarm systems. What is the probability that a warning would be sounded if a fire were to start? If he were to install three fire alarm systems, what would the corresponding probability be?

The assistant plant manager has a head on his shoulders. He points out to the manager that other users of the fire alarm system have determined that even when no fire is present, the warning will be *falsely* emitted from an individual fire alarm system 5% of the time. He then calculates that with three fire alarm systems installed, a false warning will be issued __% of the time.

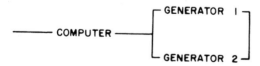

Figure 8.3 Computer system diagram.

    a. Fill in the blank.

    b. What is a crude, though useful, approximation for the percentage of interest?

4.* The assistant plant manager wishes to achieve a high assurance of receiving an immediate warning when a fire breaks out, but not at the price of constantly shutting down the plant because of *false* warnings of fire. He makes use of the three fire alarm systems, but in the following fashion:

    a. If at least two out of the three systems issue a warning that fire is present, he acts on the warning.

    b. If only one out of the three systems issues a warning that fire is present, he assumes that the bleating system is malfunctioning and that no fire is present. He continues plant operation, and has the neurotic (he hopes) system checked out.

       What is the probability that a fire will go undetected? What percent of the time will he get false warnings?

5. In a system of generators arranged in parallel, increasing the number of generators _____ the reliability of the system. Fill in the blank and explain your answer.

6. In a system consisting of subsystems arranged in series, increasing the number of subsystems _____ the reliability of the system. Fill in the blank and explain your answer.

7. Discuss the role of a statistician on a research team whose mission is to design a complex system.

8.* Refer to Problem 8 of Vignette 7. How could one define the reliability of a multistate system? How would you predict the reliability of such a system?

9. Are there some examples of systems in Vignette 8 that could be thought of as multistate systems? Explain.

10. An engineer states that a specific component has a reliability of .99. How could you investigate this claim?

Vignette 9
Round-Robin Tournaments

Statistics are the food of love.

Roger Angell, *Late Innings*

Will the panelists arm themselves with encyclopedias and fact books? "Not really," said Barber. "I probably will carry two or three record books. I would imagine, though, that my main contribution would be my memory of events and people. I think the dullest thing you can get into is statistics."

From an article by Barry Cooper, *Tallahassee Democrat*, July 31, 1981. [Cooper quoted the legendary baseball announcer Red Barber about his role in an upcoming National Public Radio call-in show called "The Great Trivia Baseball Game." The show was designed to relieve baseball-starved fans during the 1981 strike by major league baseball players.]

Table 9.1 summarizes the outcomes of the games played between baseball teams in the National League during 1965. Each of the 10 teams in the league was scheduled to play 18 games with each of the 9 other teams, or $9 \times 18 = 162$ games per team. A tournament of this type, in which each team plays every other team the *same* number of games, is called a **round-robin** tournament.

To understand Table 9.1, consider for example the entry "12" in row two, column four. This 12 signifies that San Francisco defeated Cincinnati 12 times. Similarly, the entry 6 in row four, column two signifies that Cincinnati defeated San Francisco six times during the 1965 season.

For each team the number of wins plus the number of losses must equal the number of games played by the team. Thus, for L.A. we verify that $97 + 65 = 162$, for S.F. we verify that $95 + 67 = 162$, and so on. However, for both Philadelphia and St. Louis, these sums are only 161 ($85 + 76$ for Philadelphia and $80 + 81$ for St. Louis). This is because a

Table 9.1   Competition among National League Baseball Teams, 1965[a]

	L.A.	S.F.	Pitt.	Cinc.	Milw.	Phil.	St.L.	Chic.	Hous.	N.Y.	Total Number of Wins
L.A.	—	10	9	12	10	9	12	10	13	12	97
S.F.	8	—	7	12	8	10	10	12	15	13	95
Pitt.	9	11	—	10	9	10	4	13	10	14	90
Cinc.	6	6	8	—	12	13	10	11	12	11	89
Milw.	8	10	9	6	—	6	11	9	14	13	86
Phil.	9	8	8	5	12	—	10	10	12	11	85
St.L.	6	8	14	8	7	7	—	8	9	13	80
Chic.	8	6	5	7	9	8	10	—	8	11	72
Hous.	5	3	8	6	4	6	9	10	—	14	65
N.Y.	6	5	4	7	5	7	5	7	4	—	50
Total number of losses	65	67	72	73	76	76	81	90	97	112	

[a]Number of games won by the team in the left column playing the team in the top row.

game between Philadelphia and St. Louis was cancelled because of rain, and the game was not rescheduled.

Now comes the big question: Which team should be declared the winner of the tournament? Under the rules of the National League, the team that wins the greatest number of games is the National League pennant winner. L.A., with 97 wins, is clearly the tournament winner.

But now a more subtle question arises: Is L.A. really the *best* team in the National League? The immediate rejoinder comes to mind: What do you mean by *best*?

Actually there are two quite distinct aspects to the last question: "Best" may correspond to any one of a number of *innate* characteristics:

1. Ability to win the largest number of games
2. Ability to beat each of the other nine teams in the league
3. Ability to beat more than half of the other nine teams in the league, and so on

Suppose we choose a particular one of the possible definitions of best, say definition 1. It is certainly possible for a team, say L.A., to be the best in the sense of having the greatest *innate* ability to win the largest number of games, and yet through bad luck during the season, not *actually* win the largest number of games. In other words, there is a fundamental distinction between the unknown value of interest (in the present case, the **probability** that L.A. wins the largest number of games), and the corresponding actual outcome (in the 1965 season, L.A. won 97 out of 162 games played, which *is* the largest number of wins attained among the 10 teams in the league, as shown in the last column of Table 9.1).

In fact, one of the fundamental problems tackled by statistics is to **estimate** an unknown value (in the language of statisticians, a **parameter**) from **observations** taken on this parameter. (In the present discussion, the parameter is *the probability that L.A. wins the greatest number of games*, while the observations corresponding to this parameter are the numbers of games won by the various teams in the National League.)

Under the present definition of best team, the observation that L.A. won the greatest number of games among all 10 teams would lead us to believe that L.A. may have the largest probability of winning the greatest number of games.

Suppose instead that the best team in the league is the team having the ability to beat *each* other team in the league (that is, definition 2). To measure this ability, we would take observations as shown:

	Opponent								
	S.F.	Pitt.	Cinc.	Milw.	Phil.	St.L.	Chic.	Hous.	N.Y.
Proportion of times L.A. Won Against Opponent	.556	.500	.667	.556	.500	.667	.556	.722	.667

The observations again suggest that L.A. is the best team, since the observed proportion of games won is .5 or higher against every other team in the league. Against Pittsburgh and Philadelphia, L.A. did not display a definitive superiority, since in each case L.A. lost as many games (9) as it won (in baseball jargon, L.A. and the opposing team "split" the 18 games played in each case). Thus, from the evidence available, there is no reason to believe that L.A. is actually better than Pittsburgh or that L.A. is actually better than Philadelphia.

If the second definition of best were actually used (that is, the ability to beat every other team in the league), it could easily turn out that there is *no* best team in the league. For example, suppose L.A. did have the ability to beat every other team except Pittsburgh; suppose, however, against Pittsburgh, L.A. would be likely to lose. Putting it another way, suppose that if

L.A. and Pittsburgh were to play a long series of games, Pittsburgh would win more than one-half the games. Under these assumptions, we would conclude that no one of the teams seemed best, based on the observed outcomes, since L.A. is *not* the best (Pittsburgh being better), and Pittsburgh is not the best, since it won only 4 of the 18 games played against St. Louis.

In actual practice, the team that wins the most games is declared the winner of the tournament. In other words, the choice of the tournament winner is based on the observed outcome (L.A. wins the greatest number of games) rather than on the corresponding unknown parameter value, namely, the probability that L.A. wins the most games. Obviously, this method of determining the tournament winner is easy to administer and requires no sophisticated or complicated calculations. If the National League were to attempt to estimate the probability that L.A. wins the most games with a reasonably high degree of assurance, a number of repetitions of the tournament would have to be held—an obviously unfeasible procedure, exhausting both players and fans.

(Incidentally, the method used for choosing the league winner changed after 1969. The league was divided into two divisions. At the end of the regular season, there were divisional playoffs to determine the league winner. We will not concern ourselves with this more complicated method.)

As most baseball fans know, after the regular season ends, the National League and the American League winners are pitted against each other in a best-of-seven World Series to determine the world champion. Although seven games *may* be required (delighting fans and enriching team owners), the series actually ends when one of the two competing teams first wins four games. This winning team is then pronounced world champion.

The same subtle statistical problem faces us. For example, in 1965, the World Series was played between L.A. and Minnesota. Minnesota won games 1 and 2, L.A. won games 3, 4, and 5, Minnesota won game 6, and L.A. won game 7. Thus seven games were required for the final decision, and L.A. won the series four games to three. Although we have observed that L.A. won more than half the games played, can we conclude that L.A. is necessarily the better team, that is, has a probability greater than .5 of beating Minnesota? Isn't it possible that L.A. has only a .4 probability of winning a typical game with Minnesota, but through "luck", chance, statistical fluctuations—call it what you like—L.A. wins the World Series? That is, isn't it possible that if L.A. were to play Minnesota a long sequence of games, L.A. would only win about 40% of the games played, but in a particular best-of-seven World Series, L.A. prevails?

Thus, it is important to distinguish between the probability that L.A. wins a typical game played against Minnesota and the observed proportion of games won by L.A. in a small number of games played between the two teams. In fact, we have again run into the fundamental problem of statistics — to estimate an unknown parameter (the true probability that L.A. beats Minnesota in a typical game) from an observed outcome (L.A. wins four out of the seven games played against Minnesota in the 1965 World Series).

The interested reader may wish to read Gibbons, Olkin, and Sobel (1978). They discuss the problem of deciding whether the team that does win is actually the best team in a more general context. Their basic goal is to determine the number of games that should be played in order to have a high confidence that the best team actually wins.

Summary

1. In a **round-robin tournament**, each team plays every other team the same number of games.
2. In round-robin tournaments the usual practice is to declare as winner the team that wins the most games. Reasonable alternative ways of choosing a winner exist.
3. The team with the highest probability of winning the greatest number of games may not actually win the most games.

Problems

1. Under the criterion that "best" corresponds to the ability to beat every other team in the league, may there be more than one best team? Explain.
2. Answer the same question using the criterion that "best" corresponds to the ability to beat more than half of the other teams in the league.
3. Wilton Grimes has his own scheme for picking the winner of a round-robin tournament.

   a. First, he lists each team in the tournament. For the 1965 National League tournament, he would list the teams in the first column:

L.A.:	13
S.F.:	15
Pitt.:	14
Cinc.:	13
Milw.:	14
Phil.:	12
St.L.:	14
Chi.:	11
Hous.:	14
N.Y.:	7

Table 9.2 Competition among American League Baseball Teams, 1965[a]

	Minn.	Chic.	Balt.	Detr.	Clev.	N.Y.	Cali.	Wash.	Bost.	K.C.	Total Number of Wins
Minn.	—	11	10	10	7	13	9	15	17	10	102
Chic.	7	—	9	9	10	8	12	13	14	13	95
Balt.	8	9	—	11	10	13	13	8	11	11	94
Detr.	8	9	7	—	9	10	10	11	12	13	89
Clev.	11	8	8	9	—	12	9	11	10	9	87
N.Y.	5	10	5	8	6	—	12	11	9	11	77
Cali.	9	6	5	8	9	6	—	6	13	13	75
Wash.	3	5	10	7	7	7	12	—	7	12	70
Bost.	1	4	7	6	8	9	5	11	—	11	62
K.C.	8	5	7	5	9	7	5	6	7	—	59
Total number of losses	60	67	68	73	75	85	87	92	100	103	

[a]Number of games won by the team in the left column when playing the team in the top row.

b. Next, he enters the greatest number of games won by L.A. in playing each of its nine opponents in the league—this happens to be 13 wins against Houston. Similarly, San Francisco won the greatest number, 15, of games against Houston, among its nine opponents in the league.

c. Finally, he declares S.F. the 1965 National League pennant winner since S.F. leads the list with 15.

What are the defects of Grimes' method?

Table 9.2 lists the results of the 1965 American League baseball tournament. Use Grimes' method to determine the American League champion.

4. Devise your own scheme for selecting the best team in a round-robin tournament. What are its advantages and disadvantages?

5. Suppose Philadelphia and Milwaukee had played two additional games in the 1965 National League tournament. Use Table 9.1 to estimate the probability that Philadelphia would have won both games. What assumptions are you making?

6. How would you use the information of Tables 9.1 and 9.2 to estimate:
   a. The probability that L.A. beats Minn. in a typical game
   b. The probability that L.A. wins a World Series against Minn.
   Explain your answers.

7. Table 9.1 shows that L.A. won 97 games of 162 played, representing a winning proportion of .599. Table 9.2 shows that Minnesota won 102 games out of 162 played, a winning proportion of .630. Can these winning proportions be used to predict with reasonable accuracy the winner of the World Series? What additional data would be helpful?

8. What are reasonable definitions of the *worst* team in a round-robin tournament?

9. Describe round-robin competitions in which a team winning the greatest number of games is not certain of being declared the winner.

10. Is the winner of a "knock-out" tournament always the best team in the tournament? Explain. (We will not define a knock-out tournament here. If you do not know what it is, make an educated guess.)[†]

11.* Table 9.1 does not give the order in which the teams played their opponents. Can such information be used to determine which team is best? (Hint: There may be carryover effects. For example, if L.A. plays at S.F. on Sunday and then at N.Y. on Monday, L.A.'s performance against N.Y. may have been affected by the S.F. game and travel.)

12.* Refer to Problem 11. How would you design a round-robin tournament to balance carryover effects?

Reference

1. J. D. Gibbons, I. Olkin, and M. Sobel (1978). Baseball competitions—Are enough games played? *The American Statistician* 32(3), 89–95.

Reference 1 is fun to read, especially if you are a baseball fan. For example, one of the conclusions in reference 1 is that the best of seven-game World Series format does not contain enough games to determine with high probability the best team. Many frustrated fans have already reached that conclusion.

[†] [M. H. formulated this question. (1) I think M. H. means "*un*educated guess." (2) I don't see why M. H. does not either explain "knock-out tournament" or knock out the question. F. P.]

SAMPLING

An utterly steady, reliable woman, responsible to the point of grimness. Daisy was a statistician for the Gallup Poll.

Saul Bellow, *Herzog*

"They claim jury selection was fair and statistically sound! All prospective jurors were obtained from a *random sample* of the people working at Wright-Patterson Air Force Base."

Vignette 10
Estimating a Population Property from a Sample

One of the central problems in statistics is to estimate some property of a population from a sample selected from the population. Let's consider some examples.

1. The population consists of full-time workers in the United States. The property is the average weekly income. The sample consists of 1,000 full-time workers randomly selected by Social Security number.
2. The population consists of Boeing Co. employees. The property is the average distance from home to place of employment.
3. The population consists of blue whales throughout the Antarctic. The property is the number of blue whales. The sample consists of blue whales tagged after their first capture that are then recaptured (see Vignette 14, "Capture-Recapture").
4. The population consists of registered voters. The property is the proportion of voters who will vote for the Democratic candidate for president. The sample consists of 5,000 registered voters skillfully selected to reflect various categories present in the voting population. These categories could include geographical, racial, ethnic, economic and political categories.
5. The population consists of an unending sequence of tosses of a nickel. The property is the long-run proportion of outcomes in which a head is showing. The sample consists of 100 tosses of the nickel. Note that the population in this case is *conceptual* rather than actual, as were the preceding four populations.

A small philosophical point: Conceptual populations may be infinite in size, that is, the number of members is unending. Actual populations are finite, although in some cases, the number of members may be very large. Strangely enough, infinite-size populations are usually easier to analyze mathematically than are finite-size populations. (This may confirm the reader's opinion that mathematicians constitute a population with the property of being strange; this opinion may be based on a sample of mathematicians that the reader has met or taken courses from.)

Clearly, there are several key aspects to the problem of reaching valid conclusions about the population from a sample.

First, the population must be clearly defined. Thus, in example 1, does the population consist of full-time workers only, or does it also include part-time workers? Does it include seasonal workers (such as crop pickers), or not? Does it include full-time temporary workers such as students during the summer break?

Second, the population property of interest must be precisely specified. In example 1, the property of interest is stated as the average weekly income. The word "average" most commonly is used for the sum of the values divided by the number of values (**arithmetic mean**). However, average is a broader concept than just the arithmetic mean. It also refers to the **geometric mean**. To find the geometric mean of a set of values find a number which when multiplied by itself as many times as there are elements in the set will yield the product of the elements in the set. For example, the geometric mean of (3, 6, 12) is 6, since $6 \times 6 \times 6 = 3 \times 6 \times 12$. Alternatively, the **median** of the population qualifies as a population average. The median of a set of an odd number of values is the middle one of the values arranged in order of size. For example, the median of (1, 10, 1000) is 10. The median of an even number of values is the arithmetic mean of the middle two values. For example, the median of (1, 10, 20, 1000) is $(10 + 20)/2 = 15$.

Finally, the sample must be carefully drawn from the population so that it is likely to be a **representative sample** *of the population.* By this we mean that we can reach a conclusion from observing only the sample, which is reasonably accurate for the population as a whole. [But see Kruskal and Mosteller (1979) for how the expression "representative sample" is used and misused in various ways in the statistical literature.] Thus, from a skillfully drawn sample of full-time employees in the United States, we might find that the mean income in the sample is $10,300. If we had gone to the trouble of determining the income of each full-time employee in the United States, we might find the population mean income is $10,500. The **error** in relying on the sample estimate is not even 2% in this case.

Drawing the sample so that it is representative of the population is not an easy task in many practical problems. Several alternative methods are available:

Simple Random Sample. Every member of the population has an equal chance of being drawn. Often a table of random numbers is used to achieve this requirement efficiently and objectively. Specifically, the members of the population are assigned numbers in succession: 1, 2, 3, ... Then from a table of random numbers, 100 numbers are selected (if the sample is to be of size 100). The corresponding members of the population are then selected, constituting the sample (see Vignette 12, "Using a Table of Random Numbers").

Stratified Sample. In example 4, the value of a stratified sample is most apparent. We know that different races have different political preferences. Thus, we would like to ensure that the sample contains members of the different races. Similarly, we know that different geographical regions display different political preferences. Again, we want to make sure that the sample contains registered voters from the different geographical regions. Similar considerations apply to all the other *strata* that enter into the population of registered voters. Thus, a stratified sample is a sample that is selected to contain representatives of the various strata present in the population as a whole.

An important question now arises. How many members of a given stratum should be selected for the sample? One obvious (though incorrect) answer is to select the number of representatives from each stratum in proportion to the relative size of the stratum. Thus, if 7% of all registered voters are New Yorkers, we would require that 7% of our sample consist of New Yorkers. In practice, other considerations enter into the determination of the number to be selected from each stratum. A key consideration is illustrated by the following reasoning.

Suppose, as a hypothetical example, blacks tend to vote as a group. Clearly, it would suffice to take a smaller number of blacks for the sample than suggested by their proportion among registered voters. Suppose, as a second hypothetical example, that young voters were quite unpredictable in their voting habits. We would be wise to include more young people in our sample than would be suggested by their proportion in the population. These two examples illustrate the general principle that the less predictable the voting habits of a group, the greater should be the representation of that group in the sample. Thus, the determination of the composition of the sample depends on several factors, such as the size of the stratum, the heterogeneity among the members of the stratum, and so forth.

A good stratified sample is clearly difficult to construct. Think of the great variety of refined strata that exist. Do we want to ensure that the proportion of black female union members living in New York who are registered to vote are correctly represented in the sample? You can see the problems of trying to take into account all possible combinations of factors.

In spite of these difficulties, properly constructed stratified samples are quite useful in estimating population properties.

The next important question facing us is: What size sample should we take? It is *not* true that the sample size should be proportional to the population size. (If it were, we would be required to take an infinite-size sample in example 5, which, of course, is silly.) Statisticians have developed methods to calculate the appropriate sample size depending on the type of population, the variability among members of the population, the precision needed for the estimate, the cost of sampling and observation, and other such factors.

Let us examine more closely the factor mentioned previously, "the precision needed for the estimate." Note first that it implicitly is telling us that the estimate will generally differ somewhat from the true population value. In example 5, suppose that the long-run proportion of heads in an unending sequence of tosses of a nickel is .5. In tossing the nickel 100 times, we might end up with 55 heads, differing by .05 from the long-run proportion of heads. Thus, if we were to estimate the long-run proportion of heads by .55, the observed proportion of heads, we would experience an error of .05. Common sense (as well as statistical theory) tells us that the larger the sample we take, the smaller will the error tend to be. Thus, if we had tossed the nickel a wearisome 1,000 times, the chance of observing 55% or more heads would be less than .001, a very small chance indeed.

Summary

1. An important problem in statistics is to estimate one or more properties of a population from a sample deliberately chosen to be representative of the population.
2. Both the population and the property have to be well defined for correct estimation.
3. Several methods are used to choose a representative sample.
   a. A random sample gives every member of the population an equal chance to be selected for the sample.
   b. A stratified sample is selected taking into account the strata within the population.
   A table of random numbers may be used in sampling to avoid bias.
4. The size of the sample depends on the size of the population (but not at all in direct proportion), the complexity of strata in the population, the degree of precision required, and the cost of sampling and observation, among other factors.

Problems

1. A bank executive wishes to determine the average amount of money held in a savings account in his bank. Describe in detail a procedure for estimating this amount without examining every savings account.

2. The bank executive becomes more curious about the savings account operation after he completes the study in Problem 1. He now wishes to estimate the average number of deposits made per account each year. He sets up a sampling plan after a three-martini lunch with one of his biggest[†] loan customers. His plan is to:

    a. Select every tenth deposit made in the month of January.

    b. Determine the corresponding account number.

    c. Sort these account numbers to arrive at a list consisting of distinct account numbers and the number of deposits made into the corresponding accounts.

    d. Finally, obtain the average of the numbers of deposits described in c.

    Describe each fallacy or weakness in his plan. How would *you* set up a sampling plan to obtain the information he wants?

3. A criminology professor wishes to determine the proportion of students at Florida State University who are currently smoking marijuana. He selects a sample of 500 students with the aid of a table of random numbers. He mails a simple questionnaire to each with questions concerning the smoking of marijuana, assuring the recipients that their replies will be treated with full confidentiality; to emphasize this point, he provides unmarked return envelopes.

    Only 65% of the questionnaires are returned with the information requested. The professor then summarizes the information provided.

    How valid are the professor's conclusions? What is the chief source of bias in the procedure followed? What additional steps will improve the reliability of the conclusions?

4. Toss a nickel 100 times in succession. Plot the proportion of heads versus the toss number (1 for the first toss, 100 for the last toss).

    a. Do the fluctuations tend to get smaller?

    b. What is your estimate of the proportion of heads that would show up in the long run?

    c. Estimate the long-run proportion of heads from the first 50 tosses, and then separately from the second 50 tosses. How far apart are

---

[†] Seven feet, four inches, a retired basketball player.

the two estimates? If the number of tosses had been 1,000 instead of 100, and estimates had been separately obtained from the two halves of the sample, would the discrepancy tend to be smaller?

d. Average the two estimates obtained in c. How does the average compare with the estimate obtained in b? Why?

5. A poorly trained teacher of statistics tosses one nickel 1,000 times, obtaining heads in 507 of the trials. He then makes a claim concerning the long-run proportion of heads that will show up in tossing a nickel, randomly selected, and also adds an allowance for error since his estimate is based on a finite number of trials. What is the fallacy of his reasoning in putting forward his claim? How should he have obtained data to make the type of claim he wishes to make?

6. A pollster wishes to predict the winner of a mayoralty election. He has his staff call 2,000 telephone numbers randomly selected from the city phone book and ask for the preference of the telephone respondent. His staff works from 9 a.m.-5 p.m.

What factors are likely to bias his prediction? How can he improve the validity of his conclusion without increasing the sample size?[†]

7. A new intra-uterine device is designed to prevent pregnancy. To study its effectiveness, what population would be of interest? What property of the population would one want to estimate?

8. A university faculty is considering forming a union. How could you assess the degree of support for a union without polling the entire faculty?

9.* Even with careful planning, there exists the risk that the sample obtained may *not* be representative of the population. Explain.

10.* Can you describe a situation where it would be *unwise* to fix the size of the sample in advance?

Reference

1. W. Kruskal and F. Mosteller (1979). Representative sampling, III: the current statistical literature. *International Statistical Review 47*, 245–265.

Reference 1 discusses and classifies various meanings of the terms "representative sample" and "representative sampling" as they are used in the current statistical literature. It also highlights some inconsistencies in the use of these terms by certain writers.

[†] Make sure his staff calls the city in which the election is to be held rather than the city in which the pollster and his staff live—3,000 miles away. The pollster is calling residents of his own city because it is much cheaper.

Vignette 11

Selecting a Sample from a Population: The *Literary Digest's* Fatal Indigestion

Wirthlin is a nice fellow, as straight as the borders of his native Utah, but he is the president's pollster, and presidents should not have pollsters. Presidents pay too much attention to evanescent opinion.

From an article, "Place Not Your Faith in Polls," by George F. Will, *Newsweek*, June 6, 1983

As explained in Vignette 10, a basic problem in statistics is to reach conclusions concerning a population of items from observations made on a sample of items selected from the population. A key question is, How should the sample be selected from the population?

An interesting historical episode illustrates the pitfalls of poor or incorrect selection of a sample from a population. In late 1935, the *Literary Digest*, a prestigious and popular magazine of news and commentary (like *Time* or *Newsweek* these days), conducted a survey to predict whether the incumbent president, Franklin D. Roosevelt, or the challenger, Senator Alfred M. Landon, would win the 1936 presidential election. Ten million prospective voters were sent questionnaires to determine their preferences. Unfortunately, only 2.3 million of the 10 million questionnaires were returned. Based on the responses so obtained, the *Literary Digest* confidently predicted that Landon would win by a crushing 3:2 ratio.

The prediction turned out to be a disaster. Roosevelt beat Landon in a landslide victory, winning 62% of the popular vote and carrying 46 of the 48 states. Although the loss to Landon and the Republican party was serious, by contrast, the consequences to the *Literary Digest* were shattering. The magazine became a laughing stock throughout the nation and, indeed, throughout the world. Countless cartoons in newspapers and magazines lampooned the unfortunate magazine; radio commentators and comics alike devised endless variations on the "accuracy" of the *Literary Digest's*

DOONESBURY   by Garry Trudeau

DOONESBURY   by Garry Trudeau

DOONESBURY   by Garry Trudeau

Copyright, 1980, G. B. Trudeau. Reprinted with permission of Universal Press Syndicate. All rights reserved.

prediction. Within short order, the *Literary Digest*, having lost not only its former high prestige, but also a good many of its subscribers, gave up further publication and disappeared from the national scene, never to return.

A number of explanations have been offered for the terrible inaccuracy of the *Literary Digest's* prediction, most of them inaccurate themselves (see Bryson, 1976). Apparently, the real culprit (although probably not the only one) was the fact that the 23% of the recipients of the questionnaire who responded felt quite differently about the presidential contenders than did the 77% who did *not* respond. Clearly, the respondents felt strongly enough about the outcome of the election to mail in their replies, whereas the nonrespondents did not. Furthermore, the prospective voters who did feel strongly enough to respond to the *Digest's* questionnaire generally favored Landon. This, of course, led to a highly biased sample upon which the *Literary Digest* prediction was based.

The moral of this sad tale is clear: To reach a valid conclusion concerning an attribute of a population, make sure the sample selected is representative of the population. A simple way to achieve this is to select your sample so that every member of the population has the same chance of appearing in the sample. Statisticians have developed more sophisticated methods of selecting samples called **stratified random sampling** which take into account the variability in identifiable subpopulations (say, ethnic groups) of the population as a whole; their aim is to obtain an equally precise representative sample, but of smaller size.

An almost sure way to violate this principle is to ignore nonresponse in conducting a questionnaire, as demonstrated so dramatically in the *Literary Digest* case. Other ways of obtaining *nonrepresentative* samples, some obvious, some not so obvious, are as follow:

1. Select specimens for the sample possessing characteristics related to the characteristic under study. To take an extreme case, determine the average grade point average among seniors by selecting 25 Phi Beta Kappa members.
2. Select specimens from the population in a "regular," orderly, or convenient fashion. Extreme cases to dramatize the pitfalls are:
   a. Select every other person among the diners at a dinner party (probably, the hostess has deliberately arranged that men and women alternate).
   b. Select the top layer of apples in a large package of apples at the supermarket (undoubtedly, the produce handler has put the best apples where they are most visible—on top).
   c. As a college professor, you select a sample among college students to predict the outcome of a presidential election.

In Vignette 12, we will discuss the use of tables of random numbers to draw unbiased samples, that is, samples representative of the population under study.

Warning to the Reader. A number of explanations have appeared in print for the *Literary Digest's* disastrous prediction. See Gallup (1972), Huff (1954), Mendenhall, Ott, and Schaeffer (1971), and Reichard (1974). According to Bryson (1976), these explanations have one feature in common—they are all wrong. Our explanation is based on his scholarly study of the various myths that have evolved attempting to explain the *Literary Digest* failure.

Summary

1. Statistics can be used to draw conclusions concerning a population of items from observations made on a sample of items selected from the population.
2. To reach valid conclusions, the sample selected should be representative of the population.
3. In certain types of polls, nonresponse may be a significant factor leading to a nonrepresentative sample.
4. Vignette 12 shows how to draw samples representative of the population under study.

Problems

1. A political pollster wishes to predict the outcome of an election for governor. He selects a random sample from among automobile owners using the state records. Will he get a representative sample of the population of voters, or will his sample be biased? Explain your answer.

   How would *you* go about getting a representative sample?

2. A rather conservative nationally syndicated columnist wishes to determine how Americans feel about the Panama Canal Treaty (pretend the time is 1979). He asks his readers to write to him expressing their opinions. Will he be able to estimate reasonably well how Americans feel about the treaty? What sources of bias may affect his estimate?

3. A large food processing company wishes to determine which of several variations of a candy bar is most likely to appeal to the consumer. The

company asks customers in supermarkets to taste each of the several variations and state their preferences.

   a. Who will actually be more likely to eat the candy bar—children or adults?

   b. Who is more likely to be shopping at the supermarket—children or adults?

   c. Is the company obtaining a representative sample from the relevant population?

4. A congressman wishes to represent his constituency's wishes on pending gun-control legislation in the House. Upon examining his mail, he finds 1,000 letters against gun control and 800 letters in favor of it.

   a. Are the people who take the time and trouble to write representative of the population of voters in his district?

   b. Would he be correct in assuming the majority of his constituency are against gun control?

   c. Can you suggest a more scientific way of getting a representative sample of opinion of the population of voters in his district?

5. As we have learned from the sad story of the *Literary Digest* fiasco, the people who respond to a questionnaire are not necessarily representative of those who do not respond. What can a pollster do to overcome this problem of **nonresponse**?

6. In March 1980, Senator Edward Kennedy of Massachusetts soundly defeated incumbent President Jimmy Carter in the New York State Democratic Primary. Yet polls preceding the primary (including the prestigious Gallup poll) predicted that Carter would soundly defeat Kennedy in that primary. What are some possible explanations for the reversal? (Do you remember any of the postprimary explanations offered by the pollsters and news media?)

7. It has been argued that results of political polls announced before the election tend to affect the actual vote on the election day. Explain.

8. A toothpaste manufacturer sends free samples of its product to the people in a small town. It then immediately conducts a poll asking the townspeople:

   a. What is your favorite toothpaste brand?

   b. What toothpaste are you now using?

   Can such a poll be of any value? (Exclude unethical future advertising as a value.)

9. In an in-person interview, what can the interviewer do to avoid bias?

10. It is unreasonable to expect even the most carefully and scientifically conducted polls to be 100% accurate. Explain.

11. What factors affect the accuracy of a poll?

References

1. M. C. Bryson (1976). The Literary Digest: Making of a statistical myth. *The American Statistician 30*, 184–185.
2. G. Gallup (1972). Opinion polling in a democracy. In Tanur et al., *Statistics: A Guide to the Unknown.* Holden-Day, San Francisco, pp. 146–152.
3. D. Huff (1954). *How to Lie with Statistics.* Norton, New York.
4. W. Mendenhall, L. Ott, and R. L. Schaeffer (1971). *Elementary Survey Sampling.* Wadsworth, Belmont, Calif.
5. R. S. Reichard (1974). *The Figure Finaglers.* McGraw-Hill, New York.
6. *Literary Digest 122* (November 14, 1936). What went wrong with the polls? pp. 7–8.
7. R. Likert (1948). Public opinion polls. *Scientific American 179*, 7–11.

References 1–3 and 5–7 are elementary and easy to read. Reference 4 is at a slightly higher level.

Vignette 12
Using a Table of Random Numbers

We have seen in Vignette 11 (and will again see in Vignettes 16 and 20) that a table of random numbers is useful in a variety of statistical problems. In this vignette we will explain what a table of random numbers is and how it is used to generate a random sample representative of the population of interest. Finally, we will try to reassure the reader that even though particular samples selected using a table of random numbers correctly may *appear* "strange," they are just as valid as any other correctly selected sample.

What Is a Table of Random Numbers?

A table of random numbers consists of successive digits (a digit is just one of the integers 0, 1, 2 . . . , 9), usually grouped in sets of two, four, or five. Thus, a typical row in the table might appear as follows:

    0491   2117   8344   1471   6715   9620

The table might consist of as little as one page of such rows or, alternatively, might consist of a whole book containing hundreds of such pages (see, for example, Rand, 1955). (Clearly, such a book would hardly qualify for the Pulitzer Prize, though an insomniac might find it helpful.)

But what are the essential features that qualify it as a table of **random** numbers, rather than just a table of numbers? The essential features can be described as follow.

Property A. Consider any particular one-digit entry in the table (say, the ninth digit in the third row). The chance that it is a 0 is 1/10, the chance that it is a 1 is 1/10, the chance that it is a 2 is 1/10, . . . , the chance that it is a 9 is 1/10; that is, it has the same chance (**equal likelihood**) of being any of the 10 possible digits from 0 through 9.

Property B. If we know some or even all of the entries in the table except for the digit in question, the chance of the digit in question being a 0, or a 1, or a 2, and so on, *remains* 1/10; this property holds for every entry in the table. The statistician labels this property **mutual independence** among the entries.

The consequences of properties A and B are quite far-reaching. Some of these consequences are as follows.

Suppose that we are interested in numbers between 00 and 99 (two-place natural numbers). Then, starting anywhere in the table, each two-place natural number is as likely to be encountered as any other. This is also true for all successive two-place natural numbers. Thus, the first two-place natural number may be any one of 00, 01, 02, . . . , 99 with **equal likelihood**; similarly, for the second two-place natural number, the third, and so forth. Moreover, as in the case of the single digits, all of the two-place natural numbers are **mutually independent.**

In a similar fashion, each successive three-place natural number is as likely to be encountered as any other. Also, all the three-place natural numbers are mutually independent. Of course, in a similar fashion, equal likelihood and mutual independence hold for four-place natural numbers, five-place natural numbers, and so on.

At this point, the reader is no doubt wondering what all this randomness is good for. [Unless his mind has randomly wandered off, and he is thinking of his utility bill, or perhaps his lunch coming up shortly (in the sense of time, not gastric reversal.)] An example is in order at this point.

Example. An experimenter has 80 rats available. He wishes to perform an experiment to determine whether or not a certain chemical is carcinogenic. Unfortunately, his resources permit him to use only 20 rats in the experiment. How can he pick the 20 rats for the experiment at random from among the 80 available?

He starts by assigning the numbers 01, 02, 03, . . . , 80 to the individual rats in the population. (Rats hate this disregard of their individuality and consider the process derodentization.) Next, to select at random the particular 20 rats for the experiment, he enters Table A of Appendix I which contains 10,000 random digits, and picks a starting point "at random." [A philosophical point arises here. How does he ever get started at random? Must he use a second table of random numbers to find his starting point (page, line, and position) in the first table? We will gloss over this philosophical point.] For simplicity, let us say that he simply closes his eyes and points to a digit on the first page to find his starting point. Suppose this lands him on the eleventh digit of the sixth line, which happens to be 2 and continues with 5078 80729, and so on. [The random digits in Table A are grouped in five-digit numbers for convenience in reading. They could just as well have been grouped in four-digit numbers, or any handy-size numbers. The first 1,000 digits of Table A were taken from Brown and Hollander (1977, p. 415); the remaining 9,000 digits were generated by John Kitchin.]

For his purposes, the experimenter regroups the successive digits into successive pairs of digits, that is, into a sequence of two-digit numbers: 25 07 88 07 29, corresponding to the first 10 digits listed.

He now may randomly select the sample of 20 rats from among the 80 rats available. The first rat selected is #25, the second one is #07. Now he encounters a minor problem: the next two-digit number is 88, but there is no rat with such a designation. He resolves the dilemma easily: he simply discards the 88, and continues as before. Now he encounters a second type of problem. The next two-digit number is 07; but he already has included rat #07 in his sample. Again, he solves the problem easily and without biasing his random selection—he simply discards the second entry 07 encountered. [It almost looks as if we started where we did deliberately, just so we could quickly encounter the various types of problems needing resolution in using a table of random numbers. This is not so. We really did start "at random"; we just happened to be lucky (unlucky?). Incidentally "lucky" and "unlucky" are odd words to find in a statistics book.]

Continuing in this fashion—discarding two-digit numbers between 81 and 99 and two-digit numbers already included in the random sample—the experimenter develops the following worksheet:

25078 80729 27806 42877 80287 21759 61980 52447 65694 95760

Note that 88 and 98 are discarded since they lie between 81 and 99 (they are too large), while the second occurrences of 07, 80, and 28 are discarded since they have already been included in the random selection.

The randomly selected 20 two-digit numbers lying between 01 and 80, in the order of selection, are thus:

25	07	29	27	80	64	28	77	72	17
59	61	05	24	47	65	69	49	57	60

Arranging these numbers in order of size for convenience, the experimenter finally may list his random sample of 20 rat numbers as

05	07	17	24	25	27	28	29	47	49
57	59	60	61	64	65	69	72	77	80

The experimenter may now correctly claim that the 20 rats randomly selected for the experiment are typical of the population of 80 rats he has available and that no bias of any kind entered in the selection. For example, he did not deliberately or unconsciously select healthier rats, or larger rats, or meaner looking rats for the experiment. Rather, every one of the 80 rats had an equal chance of being selected for the experiment (a questionable distinction from the rat's point of view). A method such as this, which gives all possible sets of 20 rats an equal chance of being selected, is called **simple random sampling** (recall Vignette 10).

Some curious features of the final selection are noticeable.

1. Of the twenty numbers, six end in 7 (almost one-third). By chance, we would expect only about 1 in 10 such numbers.
2. None of the twenty numbers begins with a 3. The chance of such an outcome is $(7/8)^{20} = .069$, only.
3. Of the twenty numbers, fourteen are odd numbers (almost three-fourths). By chance, we would expect about one-half to be odd numbers.

Actually, such idiosyncracies of a set of randomly selected numbers are typical, rather than unusual. The important point to keep in mind is that although all possible sets of 20 distinct two-place numbers are equally likely beforehand to end up as the random sample (the method of selection ensures this), *after* any particular random selection has been made, it is very likely to display seemingly unusual, erratic, peculiar, unlikely, or biased features. The reason, of course, is that there are an almost endless number of ways in which a selection of numbers may appear unusual, erratic, peculiar, unlikely, or biased, *after the fact* (that is, after the selection has been made). Thus the striking particular random selection 01 02 03 04 · · · 19 20 is just as likely as any other random selection.

Summary

1. A table of random numbers is very useful for selecting a representative sample from a population.
2. The essential features of a table of random numbers are:
   A. The equal-likelihood feature—the chance that any particular one-digit entry is a 0 is 1/10, the chance that it is a 1 is 1/10, and so forth.
   B. The mutual independence feature—knowing some or all of the entries in the table except for the digit in question does not change the chances affecting the digit in question, that is, the chance that it is a 0 remains 1/10, the chance that it is a 1 remains 1/10, and so on.
3. When all possible samples of a given size are equally likely to be selected, such sampling is called **simple random sampling**.

Problems

1. Using Table A of Appendix I, select at random 10 even two-place numbers.

2. In the example of this vignette, we selected 20 rats at random from among 80. Now randomly divide the 20 rats chosen for the experiment into two equal-size groups of 10 each—one group to receive the alleged carcinogen, the other to constitute the control group.

3. Do the telephone numbers in a typical telephone book qualify as a table of random numbers? How would you begin checking?

4. Using the table of random numbers, select at random 50 two-place numbers from the set of 100 two-place numbers 00, 01, 02, . . . , 99.

5. Using Table A, simulate the sequence of heads and tails resulting from tossing a fair coin 100 times in succession. Note that the use of the table saves a great deal of time without loss of fidelity.

6. Why should you avoid starting at the same spot in the table of random numbers in successive uses of the table?

7. Suppose we randomly permuted the digits in a table of random numbers (perhaps by using a separate table of random numbers). Would the resulting table qualify as a table of random numbers?

8. Suppose an Israeli (or anybody else, for that matter) used the table, but read from right to left. Would the results be as valid as in reading from left to right? Carrying the idea further, suppose he read successive entries of the table vertically instead of horizontally. Would his results be equally valid?

9. Describe a table of digits in which each of the 10 digits 0, 1, 2, . . . , 9 appears equally often (10% of the time), and yet the table would not qualify as a table of random numbers.

References

1. B. W. Brown, Jr., and M. Hollander (1977). *Statistics: A Biomedical Introduction.* Wiley, New York.
2. *A Million Random Digits with 100,000 Normal Deviates* (1955).[†] The Rand Corporation. Free Press of Glencoe, Glencoe, Ill.

Reference 1 utilizes symbols, but presents biomedical applications of statistics in an elementary format. Reference 2 is useful for a statistician to own, but is much less interesting to read than a telephone directory.

[†]Normal deviates? It sounds self-contradictory.

Vignette 13

The Racing Car Problem

There was nothing to do on the train ride back to Paris except mull over the dreary weekend. The race had been uneventful. Miraculously, there were no accidents despite heavy rains which took their toll on the cars, drivers, track, and spectators. The grounds were soaked, debris was everywhere, and overnight campers had trudged around in damp dishevelment.

Claudine knew that the American would return after a two-week sulk, his usual variation on postpartum depression. His days of driving were ending, but he could still drive her crazy.

Mark Hampton, *Summer Lapse*

The following problem as described may sound a bit artificial, but it is a simplified version of many real-life problems.

You arrive late at a race track; the cars are already hurtling past you. You know from previous visits that the cars are numbered (and labeled) 1, 2, 3, . . . , up to an integer equal to the total number of cars in the race. Unfortunately, this total number is unknown to you because of your stupid late arrival.

You wish to estimate this unknown total number of cars in the race based solely on the information obtained by observing the numbers on the cars that have just streaked by you, say cars labeled 2 and 3.

Clearly, there are many possible estimators that you can use. You happen to be a very conservative type of person. (You never bet more than $5 with your bookie on the race; study all the facts about the racing cars and their drivers before you telephone your bookie; cling to false hopes for days after the race, hoping that a reversal of the observed finish will be made because of some infraction subsequently discovered and that your pick will miraculously be awarded first place; and drive home after the race obeying all traffic signs and laws whether you have won a handsome $10 or lost a heartbreaking $5.)

Thus, you pick as your estimator of the total number of cars in the race the maximum of the car numbers that you have actually observed, namely, maximum $(2, 3) = 3$. What can we say about this maximum estimator?

1. It certainly cannot **overestimate** the unknown true total number of cars in the race. If you are lucky, you may have estimated perfectly—there may be exactly 3 cars in the race. Finally, you may have estimated too low—the total number may be greater than 3. It seems clear (when an author writes this phrase, he means: It is clear to me who have spent all my life studying this subject; if it's not clear to you, dear reader, why don't you pretend it is—but, let's get on with it) that on the average, the maximum estimator will tend to **underestimate** the unknown true value.

2. On the average, how much too low is this maximum estimator? It seems clear that the size of the underestimate must depend on the unknown true total number of cars. We can reason as follows.

Suppose there are 5 cars in the race. Suppose all pairs are equally likely to be the pair actually seen. (Strictly speaking, this is not true. Why not? Hint: a car going 180,000 miles a second will pass your viewing point with higher probability than a car going at a conventional speed. But the assumption that all pairs are equally likely to be the pair actually seen is true if the numbers are assigned to the cars independently of speed.)

We now list systematically the 10 possible pairs that could have been seen, assuming 2 cars are seen, and the corresponding maximum estimator. The possible pairs are in the first column below and the corresponding maximum estimator is in the second column.

Small Digression. We see from the systematic tabulation, that there are 10 different choices of a pair of distinct numbers selected from the set $\{1, 2, 3, 4, 5\}$. For convenience we use the symbol $\binom{5}{2}$ for the number of different

Pair	Value of Maximum Estimator
1,2	2
1,3	3
1,4	4
1,5	5
2,3	3
2,4	4
2,5	5
3,4	4
3,5	5
4,5	5

choices of a pair of distinct objects chosen from among 5 distinct objects. A popular, easy-to-remember name for this symbol is "5 choose 2." It is one of a general class of such symbols called more formally **binomial coefficients** since they arise quite naturally in the binomial expansion studied in algebra. Table B in Appendix I tabulates binomial coefficients likely to be needed in modest-size problems.

We confess that we have violated our promise not to use symbols. We introduced a symbol at this point just to remind you gently that you are fortunate indeed to face only one algebraic symbol in the whole book, as compared with the standard statistics text which bristles with such symbols!

Back to our discussion of the maximum estimator: Assuming all pairs equally likely, we may next compute the average value of the maximum estimator. We simply sum the 10 possible equally likely values of the maximum estimator listed in the second column and divide by 10. Thus, we get:

$$\frac{2 + 3 + 4 + 5 + 3 + 4 + 5 + 4 + 5 + 5}{10} = \frac{40}{10} = 4.$$

Note that this average value 4 of the population of maximum estimators is an underestimate of the actual number 5 of cars in the race. In statistical lingo, we say that the estimator is **biased** low. This is not surprising since we are averaging numbers, each of which is less than or equal to the actual number 5 of cars in the race.

We can adjust the maximum estimator so that the bias is removed. The adjusted estimator is given by the following formula:

Adjusted estimator

$$= (\text{maximum estimator}) \times \left(1 + \frac{1}{\text{number of cars observed}}\right) - 1.$$

(1)

In the present example, recall that the maximum estimator is 3 (based on the observation of cars 2 and 3). Substituting 3 for the maximum estimator and 2 for the number of cars observed into formula (1), we compute:

$$\text{Adjusted estimator} = 3 \times \left(1 + \frac{1}{2}\right) - 1$$

$$= 3 \times \frac{3}{2} - 1 = 4.5 - 1 = 3.5.$$

Strange? Yes. Stupid? No. Remember our aim is to use an estimator of the number of cars in the race which *on the average* is equal to the unknown true number of cars in the race. This means that sometimes the estimator will underestimate, sometimes it will overestimate, and sometimes it will exactly estimate the unknown true value. The "sometimes" refers to the fact that we visualize using the adjusted estimator in a sequence of repetitions of the situation.

There is something weird in visualizing a sequence of repetitions. The auto race fan arrives late at each race, sees a certain number of race cars whip by, and then uses formula (1) to estimate the unknown number of cars competing in the race. We might justifiably ask:

1. Why doesn't the fan start out for the race track earlier? Can't he learn?
2. Why doesn't he get a program of forthcoming races and simply read off the number of cars in each race?
3. If this is expecting too much of him, why doesn't he ask his neighbor(s) at the race what's going on?
4. A racing car devotee showing such consistent talent at repeating the same mistake would undoubtedly bet on the race, consistently lose, and run out of money. Thus, he would be forced to end the sequence of repeated stupidity. We would never be able to find out (empirically) whether his adjusted estimator of the number of cars in the race is unbiased.

Back to the adjusted estimator! For the 10 possible pairs of cars observed, we have:

Sample	Value of Adjusted Estimator
1,2	$(3/2) \times 2 - 1 = 2$
1,3	$(3/2) \times 3 - 1 = 7/2$
1,4	$(3/2) \times 4 - 1 = 5$
1,5	$(3/2) \times 5 - 1 = 13/2$
2,3	$(3/2) \times 3 - 1 = 7/2$
2,4	$(3/2) \times 4 - 1 = 5$
2,5	$(3/2) \times 5 - 1 = 13/2$
3,4	$(3/2) \times 4 - 1 = 5$
3,5	$(3/2) \times 5 - 1 = 13/2$
4,5	$(3/2) \times 5 - 1 = 13/2$

We calculate the average value of the adjusted estimator just as we calculated the average value of the maximum estimator:

Average value of the adjusted estimator

$$= \frac{2 + (7/2) + 5 + (13/2) + (7/2) + 5 + (13/2) + 5 + (13/2) + (13/2)}{10} = 5.$$

We are using the fact that all 10 pairs of numbers $(1, 2), \ldots, (4, 5)$ are equally likely to be observed.

Note that the average value 5 is exactly equal to the true number of cars in the race; that is, the adjusted estimator is indeed **unbiased**. This calculation tends to corroborate the claim that the adjustment removes the bias.

Next in Fig. 13.1 we display the **distribution** of the adjusted estimator for our example in which the sample size is 2 and the number of cars in the race is 5: 4/10 of the time the estimator is 6½, overestimating the true value by 1½; 3/10 of the time the estimator is exactly right; 2/10 of the time the estimator is 3½, underestimating the true value 5 by 1½; and 1/10 of the time the estimator is 2, underestimating the true value 5 by 3. As we have seen, the underestimates and the overestimates just balance out to yield 0 bias *on the average*.

Naturally, we obtain a more accurate estimate of the (unknown) number of cars in the race by increasing the sample size (the number of cars actually observed). To illustrate this effect, we consider the case in which the number of cars in the race is 5, as before, but the actual number of cars observed is 3, rather than 2. We compute the number of possible samples of size 3 as $\binom{5}{3} = 10$, which happens to be the same as in the previous case. Using formula (1), we compute

$$\text{Adjusted estimator} = \frac{4}{3} \times (\text{maximum estimator}) - 1.$$

For the 10 possible samples of size 3, we obtain:

Sample	Value of Adjusted Estimator
1,2,3	3
1,2,4	13/3
1,2,5	17/3
1,3,4	13/3
1,3,5	17/3
1,4,5	17/3
2,3,4	13/3
2,3,5	17/3
2,4,5	17/3
3,4,5	17/3

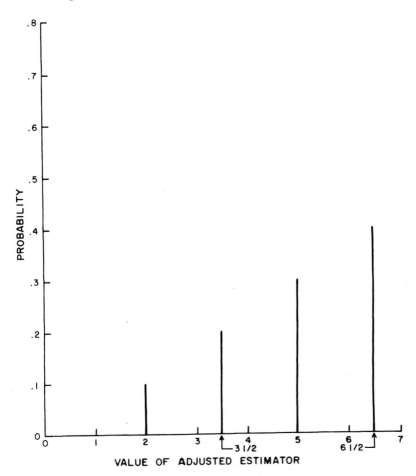

Figure 13.1   The distribution of the adjusted estimator when the sample size is 2 and the total number of cars is 5.

You will find it instructive to check that the average value of the adjusted estimator is again equal to the true value of 5. Figure 13.2 shows that for the sample size of 3, the adjusted estimator tends to be more concentrated about and thus closer to the true value of 5 than in the case in which the sample size is only 2.

From Fig. 13.2, we see that 6/10 of the time we estimate the true value 5 by 17/3, 3/10 of the time we estimate the true value 5 by 13/3, and 1/10 of the time, we estimate the true value by 3. It is easy to check that the net effect is 0 bias on the average.

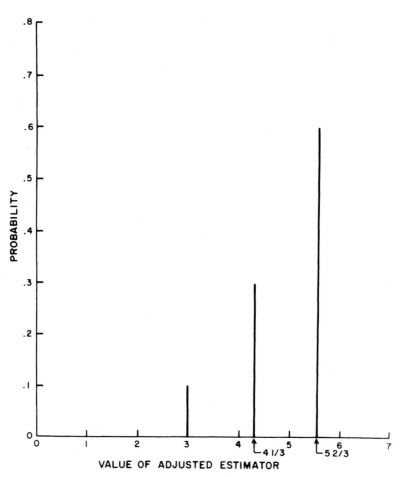

Figure 13.2   The distribution of the adjusted estimator when the sample size is 3 and the total number of cars is 5.

Supplementary Remarks.   The racing car model is not simply a charming puzzle. It has real-life applications. We sketch one in somewhat over-simplified form. An army captures enemy soldiers and notes their serial numbers. Assuming serial numbers are issued in regular sequence, the army intelligence officer may glean an estimate of the size of the opposing force. For further examples see Problems 6, 7, and 8.

"Roland, I disagree with your estimate. I estimate the number of cars in the race to be 10."

Summary

1. The racing car model assumes that we see a random sample of the total sequence of cars in a race. Our aim is to estimate the unknown total number of cars from the observed sample. The cars are numbered 1, 2, . . . , highest number.

2. A simple estimator is the largest number among the cars actually observed. This **maximum estimator** unfortunately is biased low; that is, on the average we underestimate the unknown true number of cars in the race.

3. A simple adjustment of this maximum estimator yields the **adjusted estimator**. The adjusted estimator is **unbiased**. Sometimes, it is too low,

other times too high, but on the average the net effect is 0 bias.

4. As the sample size (number of cars actually observed) increases, the adjusted estimator tends to be closer to the **parameter** (total number of cars in the race) being estimated.

Problems

1. In the racing car problem, suppose you observe a sample of size 4 and the total number of cars is 7 (although unknown to you).
   a. List the possible samples of size 4, and for each sample, calculate the value of the adjusted estimator.
   b. Verify that the average value of the adjusted estimator is 7.
   c. Plot the distribution of the adjusted estimator.
2. For the case where the sample size is 4 and the total number of cars is 7, calculate the average value of the maximum estimator.
3. Note that the maximum estimator always assumes an integer value but that the adjusted estimator can assume fractional values. Can you think of some ways to modify the adjusted estimator so that it does not assume fractional values? Bear in mind that a modification is likely to change the average value so that your new estimator may not have an average value that is equal to the total number of cars.
4. Suppose that the sample size is 3 and the total number of cars is 7.
   a. What is the chance that the adjusted estimator will overestimate the true total?
   b. What is the chance that the adjusted estimator will underestimate the true total?
   c. What is the chance that the adjusted estimator will be equal to the true total?
   d. What is the chance that the maximum estimator will be equal to the true total?
5.* Suppose that the total number of cars is 15 and the sample size is 5.
   a. What is the probability that the maximum estimator will be equal to 10?
   b. What is the probability that the maximum estimator will be equal to 11?
6. A car manufacturer that specializes in the "silver-streak" automobile assigns to successive cars produced during the year, successive integers beginning with one. This is done to enhance the status value of the car. The total number is kept a company secret, but a competitor is interested in estimating that number. By year end the competitor has seen silver-streaks with numbers 7, 28, 11. How should the competitor estimate the total silver-streak production?

7. You are walking across the campus to the duplications office in order to duplicate a book manuscript. A violent wind causes you to drop the manuscript. With pages blowing everywhere and many lost, you are able to retrieve 74 pages, with the largest page number being 187. What is your estimate of the total number of pages in the book? What is your estimate of the chance that the book publisher will ever send you another book manuscript to review? What kind of statistical pervert are you to be estimating the number of pages in the book when you should be running around like crazy trying to recover the scattering pages of the only copy of the manuscript sent to you?

8. What are some other situations to which the racing car model is applicable?

9. Skip ahead to Vignette 14. Do you see similarities between the capture-recapture problem and the racing car problem? Explain.

10.* Return to the initial situation considered at the beginning of this vignette. Suppose the sample consists of two separate observations sufficiently spaced in time so that it is possible to observe the same car twice. How would you now estimate the total number of cars in the race? What new assumptions would you want to impose?

Reference

1. A. Tenenbein (1971). The racing car problem. *The American Statistician 25*, 38–40.

Reference 1 contains a technical derivation of formula (1) which may be of interest to the more advanced reader.

Vignette 14

Capture–Recapture

NEW YORK —At least a quarter or more of the world's population of giant pandas appears to have perished as a result of a quirk of nature—the spontaneous, genetically programmed dying-out of the two species of bamboo that provide the major portion of the animal's diet. . . .

Schaller said in an interview that he expects to go to China some time next month. "We'll make a preliminary study of the area and then decide how we'll proceed from there," he said. "At some point we'll have to use radiotelemetry—attach tiny radio transmitters to the animals—to find out when and where they travel and their geographic distribution."

West Palm Beach Post, April 8, 1980

As we know only too well from newspaper headlines and TV programs, species after species of animal and fish face extinction. In some cases, remedial action can be taken to prevent this tragedy for a specific species by relocation, by halting industrial pollution, or by reducing or eliminating the hunting or fishing of the animals or fish endangered.

However, before this action can be taken, it is clearly necessary to know with reasonable precision how many fish are left and how they are distributed geographically. (For simplicity of discussion, we will use the term "fish" to refer to the endangered species, and use corresponding language to refer to its habitat and so on. Of course, a similar analysis of the extinction problem applies to animals or insects.)

Thus, we are led to the following somewhat oversimplified situation. A lake has an unknown number of fish of a dying breed. We wish to estimate this number somehow. We cannot, of course, catch every single member of this fish population and count the number caught. What shall we do?

A clever method has been developed by statisticians to estimate the unknown number of fish. Some number of fish of the dying breed are caught, say 10. A numbered tag is carefully attached to each of these fish so as not to injure it or interfere with its normal functioning. The fish is then returned to

the lake, none the worse for the experience (relieved, though puzzled). After the 10 tagged fish have had time to scatter throughout the lake, fishing is resumed, and again 10 fish of the species under study are caught. Suppose, for simplicity, 5 have tags.

We now reason as follows. Judging from the sample of fish caught, about half of the fish of the dying breed are tagged (5 out of 10). Extrapolating, the total number of fish of this species in the lake should be about 20 (that is, for every tagged fish, there is an untagged fish; recall that 10 were originally tagged).

Of course, we have simplified the problem greatly in order to make clear the main features of the solution. What complications are present in the real-life situation, not mentioned in the simple version presented above? For one thing, the real-life situation is *dynamic*—additional fish may be born during the experiment, and some of the fish, tagged or untagged, may die. In addition, it is an oversimplification to assume that each of the fish has the same chance of being caught—some fish are more gullible than others. Other factors may enter the situation. Changes in the weather may affect the chance of catching a fish. But what would complicate the situation even more would be that certain changes in the weather might affect some fish one way and other fish another way, as far as vulnerability to capture is concerned.

However, even these complications in the situation can be handled by appropriate statistical analysis. We will not try to explain these complicated models.

A few remarks are worth making concerning the capture–recapture model.

Remark 1. In most estimation problems, the statistician knows what the population consists of, but does not know what proportion of the population possesses a certain trait (for example, what proportion of the voting population will vote for the Democratic candidate for president, or as another example, what proportion of the American population will live at least 65 years). In the present estimation problem, the statistician does not know the size of the population, and wishes to estimate this unknown size.

Remark 2. The tagging method can be used to estimate properties of the population of interest other than simply size. For example, the ecologist may wish to know how the various members of the population move about. If he notes on the tag the time and location of the tagging operation, he can then determine upon recapture the distance and direction traveled by the specimen in the interval of time between tagging and recapture.

Remark 3. We have discussed the model using the term "recapture." This is most appropriate in the case of fish, since visibility underwater is generally poor. However, in the case of a population of animals, especially large animals, the recapture operation may be visual rather than physical. For example, the scientist may scan the surroundings with the aid of binoculars and count the number of specimens of the population under study that he can see, and among these, count the number bearing tags. It should be emphasized however, that for this procedure to be accurate, the determination of whether the specimen sighted bears a tag or not must be accurate.

Remark 4. We have discussed one question that can be answered by the capture-recapture method, namely, determining the size of the population. Clearly other questions can be answered by the method, such as: How much pollutant (such as lead) is present in the tissues of the average fish of the species studied? At what rate is the pollutant being accumulated? What proportion of the fish are suffering from a specified disease?

Summary

1.   It is possible to estimate the unknown size of a lake's population of fish by capturing some of the fish and tagging them, returning them to the

"No, professor, we do *not* have a capture-recapture option."

lake, recapturing (after a sufficient time lapse) a "new" sample, and noting the number of tagged fish in the recaptured sample.
2. In certain animal populations, the recapture can be visual rather than physical.

Problems

1. A scientist wishes to determine the number of fish that are 12 in. or longer in Lake Mendota (Madison, Wisconsin). He catches a number of fish from the lake, picking a variety of fishing sites throughout the

lake. He discards fish under 12 in., but tags those at least 12 in. in length, and then returns them (reluctantly) to the lake. The total number of tagged fish is 20. (He succeeds in resisting the temptation to keep one or two to show off to his fisherman friends and then eat them for dinner that evening.)

The next day he fishes again, ignoring fish under 12 in. in length, but counting 5 tagged fish among 25 fish caught of length at least 12 in. What would be his estimate of the total number of fish of length 12 in. or greater in Lake Mendota? Would the scientist be correct if he claimed that his estimate was absolutely accurate, that is, would be the same as if he had been able to count every fish of length 12 in. or greater?

2. An ecologist is conducting an experiment to determine the grazing habits of a certain species of deer. He tags 10 deer one day. Two days later he travels in his vehicle in a spiral-like route, stopping frequently to survey the surrounding territory. With the aid of a pair of powerful binoculars, he notes on a map the location and time of discovery of each tagged deer he can spot. Within 9 hours, he is fortunate enough to locate all 10 tagged deer. From the information accumulated, he makes up Table 14.1.

  a. Is the entry in the second column the *total* distance traveled by the deer in question?
  b. What is the average distance between the tag point and the discovery point?
  c. The scientist is interested in learning whether the deer travel as a

Table 14.1 Distances, Directions, and Elapsed Times of Deer Travel

Deer No.	Distance Traveled (ft)	Direction[a]	Time Elapsed (hr)
1	2,600	30	52
2	1,800	90	48
3	3,900	180	49
4	5,100	30	57
5	2,900	70	50
6	3,700	240	48
7	6,400	190	55
8	1,200	200	52
9	8,600	10	53
10	9,500	60	49

[a]Due east is reckoned as 0°, due north as 90°, due west as 180°, and due south as 270°.

"You'll get your $5 security deposit back when you return the tag!"

group or individually. What conclusions would he reach on this point?

3. What difficulties could arise with the capture–recapture estimation scheme if tagged creatures are ostracized by their untagged counterparts?

4. The estimate of the size of the population can be viewed as being obtained by equating the proportion of tagged fish in the population to the proportion of tagged fish in the recaptured sample. Explain.

5. Note the estimate of the size of the population is equal to:

$$(\text{number of tagged fish in population}) \times \frac{(\text{size of recaptured sample})}{(\text{number of tagged fish in recaptured sample})}.$$

Explain why this is intuitively reasonable.

6. What do you conclude if the recaptured sample contains only fish that are tagged?

7. Suppose there are no tagged fish in the recaptured sample. What difficulties does this present?

8. Statisticians, recognizing the dilemma hinted at in Problem 7, have devised "inverse sampling" schemes in which the recapturing continues until a specified number of tagged fish are recaptured. What are some disadvantages of such an approach?

9. What are some situations in which the capture–recapture approach provides useful information even when the size of the population is known in advance?

10. What are some possible applications of the capture–recapture approach in *human* populations?

Vignette 15

Negotiating a Contract in the Absence of Competitive Bids

WASHINGTON—The chairman of the Senate Armed Services Committee put major defense contractors on notice Thursday that he wanted an end to overly high prices for military spare parts. . . .

"We're going to get to the bottom of this sordid business of overcharges on spare parts and make sure the taxpayer gets his money's worth," Tower said. . . .

Senator Carl Levin, D-Mich., also sounded a warning against inflated spare-parts prices.

"Some of the prices we are paying for parts are nothing more than unconscionable gouging," he said.

But Levin said much of the blame must rest with the Defense Department, which sets the rules under which contractors fix their prices.

From an AP news release, October 28, 1983

The United States has a free enterprise system that (ordinarily) relies on competitive bidding. For example, a city government that wants to build a new 13,000-seat coliseum for sporting events seeks bids on the project and, subject to certain quality constraints, awards the contract to the lowest bidder. However, there are situations where billions of dollars are spent each year without the element of competition.

Wallenius (1980) describes situations in which the U.S. Department of Defense, in the process of acquiring military systems, must deal with a *sole source*. These include the situations which follow.

1. *Change orders:* It is inevitable that, after a prime contract has been awarded, design changes will need to be made *after* production begins. These changes may be the consequence of unanticipated technical problems, or may result from new performance requirements insisted upon by the U.S. government. When a design change is to be made, the original contract has to be amended, and a new price has to be negotiated between the government

111

and the contractor. Note that the government and the contractor must arrive at a new price, and the government does not have the luxury of a competitive bid at this point.

2. *Spare parts:* The number of spare parts needed for a military system is usually not known at the time the prime contract is awarded. The number of spare parts needed will depend on the reliability of the system, and such information is acquired as the production process continues. Although decisions as to the required number of spare parts (and related items such as special testing equipment, operating and repair manuals, maintenance items, and so on) come later, it is clear that the prime contractor's knowledge already gained by having done initial (and costly) engineering, tooling, and manufacturing functions for the project gives him the inside track to win any future competition. Thus, the government negotiates with the prime contractor, without the basis of competition.

3. *Modification or repair of system after production:* The builder of the system, because of experience, has the know-how and thus would win any competition for modification. In effect, there is no competition.

The government knows that it is at a severe disadvantage in these no-competition situations, and it has rules and regulations designed to minimize costs. Unfortunately, the volume of work required to carefully analyze and investigate the cost of each modification is enormous. Usually a group of government employees is assigned to study and analyze each change at the contractor's facility. The volume of work, plus a shortage of qualified government analysts, creates huge backlogs of unprocessed proposals. (Wallenius cites an aircraft plant where there were 625 unprocessed proposals totaling nearly half a billion dollars!) These backlogs sometimes force the contractor to borrow capital to cover funds tied up in the backlogs. The temptation is thus there for the government to speed up its cost analyses, but haste often leads to inaccuracy and costly overpayments.

What is needed is a method by which government analysts can estimate what the total negotiated price would be if every proposal were carefully analyzed and negotiated, but the method should not require that each proposal actually be carefully analyzed and negotiated. This can be done using **sampling**. Roughly speaking, the method is to analyze the accuracy of proposed costs and negotiate a price for a selected sample of the proposals in a backlog. Then information from the sample is extrapolated to the unsampled proposals. For example, if the average negotiated price in the sample is 80% of the average proposed price in the sample, we would estimate the negotiated price of each unsampled proposal to be 80% of its proposed price. This is known as the **ratio estimation technique**.

How should the sample of proposals be obtained? There are several meritorious possibilities. To illustrate the ideas we consider the following specific example. Suppose there is a backlog that has four proposals with proposed contract prices $100,000, $30,000, $20,000, and $10,000, and that we utilize a sample of size 2. Simple random sampling (recall Vignettes 10, 12, and 13) selects the sample so that each of the $\binom{4}{2} = 6$ possible samples has the same chance, namely 1/6, of being the sample chosen. The six possible samples of size 2 are:

($100,000, $30,000)	($30,000, $20,000)
($100,000, $20,000)	($30,000, $10,000)
($100,000, $10,000)	($20,000, $10,000).

Note that the method of simple random sampling does not use the proposed contract price in sample selection. A method of sampling that does use this auxiliary information is **probability-proportional-to-aggregate-size sampling (PPAS)**. This method of sampling requires that the probability of obtaining a given sample be proportional to the total proposed price of the sample. The aggregate prices of the six samples listed above are:

$130,000	$50,000
$120,000	$40,000
$110,000	$30,000.

The sum of these aggregate prices is $480,000. PPAS requires the probabilities of choosing the six samples to be proportional to aggregate price; thus, these probabilities must be 13/48, 12/48, 11/48, 5/48, 4/48, and 3/48, respectively. Note that PPAS reduces the probability of getting a sample with both proposals small. For example, with simple random sampling the chance of selecting a sample that does not contain the $100,000 contract is 1/2, but with PPAS this chance is reduced to $(5/48) + (4/48) + (3/48) = 1/4$.

Suppose the contractor has a tendency to *pad* the contract prices. Can the government nullify the apparent advantage to be gained by the contractor by padding? The answer is yes. We will show, via an example, that by using PPAS in conjunction with the ratio estimation technique, the contractor's *expected* advantage is zero.

Here is a simplifed example. Suppose there are only two proposals in the backlog, which should be negotiated at prices 10 and 20, respectively. The contractor properly prices the first proposal at 10 but pads the second proposal by 50% so that the contractor's price for the second proposal is 30. Suppose the government takes a sample of size 1 and negotiates both proposals on the basis of the observed sample. That is, if the first proposal is the one selected in the sample, the government finds there is no padding factor and pays the contractor's prices on both proposals, overpaying the contractor by 10. If the second proposal is the one selected in the sample, the

"Yes, General, I've just completed the cost analysis. The estimated cost for the Army's new radar system is $45,377,368."

government finds the .5 padding factor and therefore reduces the first proposal by a factor of $1/(1 + .5) = 2/3$, which results in an underaward of $10 - [10 \times (2/3)] = 10/3$ or (equivalently) an overaward of $-10/3$.

The contractor's expected overaward depends on how the sample is selected. Let us see what happens when the sample of size 1 is chosen by the government by PPAS. Although the padding factor (in our example .5) is *unknown* to the government, the proposed prices are known, and thus the total proposed price is known (in our example the total proposed price is 40). Using PPAS, the probability of selecting proposal 1 is 10/40 and the probability of selecting proposal 2 is 30/40. Thus, the expected overaward to the contractor is

(overaward if proposal 1 is chosen) $\times$ (probability proposal 1 is chosen)
+ (overaward if proposal 2 is chosen) $\times$ (probability proposal 2 is chosen)

$$= 10 \times \frac{10}{40} + \frac{-10}{3} \times \frac{30}{40}$$

$$= \frac{10}{4} - \frac{10}{4} = 0.$$

Although our illustration used the specific padding factor of .5, the ratio estimation technique in conjunction with PPAS yields an expected overaward of 0 no matter what the padding factor is!

Summary

1. Statistical sampling can be very useful in the analysis of proposals submitted to sell military systems to the Department of Defense. In particular, sampling can save the government and the contractors time and money.
2. The ratio estimation technique, in conjunction with probability-proportional-to-aggregate-size sampling, yields an expected overaward to the contractor of zero, even when the contractor pads.

Problems

1. Suppose there are only two proposals in the backlog, which should be negotiated at prices 10 and 20, respectively. The contractor properly prices the first proposal at 10 but pads the second proposal by 70%.
   a. Using PPAS to choose a sample of size 1, with what probabilities should proposals 1 and 2 be selected?
   b. Show that in this case the ratio estimation technique, in conjunction with PPAS, yields an expected overaward of zero.
2. Refer to Problem 1. Suppose the ratio estimation technique is used in conjunction with simple random sampling. Calculate the contractor's expected overaward.
3.* Again consider the situation where there are two proposals which should be negotiated at prices of 10 and 20, respectively, but suppose the contractor uses a *negative* pad of $-.7$ on the second proposal, but leaves the first proposal unchanged. Then if the first proposal is selected in the sample, no pad is found and the proposed price of $(1 - .7) \times 20$ is paid for the second proposal. If the second proposal is

selected, the negative pad is found, and the first proposal is increased by a factor of $1/(1 - .7) = 10/3$, resulting in an overaward of $70/3$. Show that if simple random sampling is used to select the sample, this negative pad of $-.7$ leads to an expected overaward that is greater than zero. Thus, the contractor can even benefit by (some) negative pads. Give an intuitive explanation of why the contractor can benefit by (some) negative pads. Under what circumstances would the government (be willing to) pay more than asked?

4.* Return to Problem 3 and find a negative padding factor for which the contractor's expected overaward is negative.

5.* Although the ratio estimation technique, in conjunction with PPAS, yields a contractor's expected overaward equal to zero, one could still be unlucky and select a sample that contained mostly large proposals or mostly small proposals. Sketch a sampling scheme that would lead to "balanced" samples (samples having small, medium-sized, and large proposals).

6. Construct an example with a backlog of three proposals, for which a sample of size 1 is to be selected. In your example, show that PPAS in conjunction with the ratio estimation technique yields an expected overaward of zero.

7. In our context of negotiating a contract in the absence of competitive bids, describe some
a. Advantages of increasing the sample size
b. Disadvantages of increasing the sample size

8. Suppose the contractor knew that the government's sampling scheme when presented with a backlog of 100 proposals is to take the following "sample" of size 10: the proposal with largest price, the proposal with the tenth largest price, the proposal with the twentieth largest price, and so forth. How could the contractor take advantage of this knowledge?

9. Suppose there are 100 proposals in the backlog and the government's sampling scheme is to take a "systematic" sample of size 10 as follows: Select a proposal at random from among the 10 largest, select a proposal at random from among the next 10 largest, and so forth. What are some advantages and disadvantages of this approach?

10. Describe some other situations when it may be necessary to negotiate in the absence of competition.

Reference

1. K. T. Wallenius (1980). Statistical methods in sole source contract negotiation. *The Journal of Undergraduate Mathematics and Its Applications 1,* 35–47.

It would seem appropriate in the spirit of competition (the theme of this vignette) to have a second reference describing how the contractor should propose bids to outwit the government. However, we are not aware of such references. This might mean (1) if the contractor does have a scheme to overcome the methods described in Wallenius' paper, he is not stupid enough to publish it in the open literature for his competitors or the government to see, or (2) the contractor has no such scheme, but is pretending that (1) holds.

LEARNING FROM DATA

I don't believe you. Not because you're a poor liar, but because it doesn't conform with the facts. I work with statistics, Mr. Washburn, or Mr. Bourne, or whatever your name is. I respect observable data and I can spot inaccuracies; I'm trained to do that.

Robert Ludlum, The Bourne Identity

"Madam, I misunderstood your advertisement. I thought you were looking for an experienced practitioner of sadistics."

Vignette 16
Was the 1970 Draft Lottery Fair?

On December 1, 1969, the U.S. Selective Service performed a lottery based on birthdates to determine the 1970 draft. The method of drawing the birthdates is described in the following account which appeared in the *New York Times* on January 4, 1970; unfortunately, no official detailed description is available, according to a Stephen Fienberg article in *Science* (1971).

Over the weekend before the December 1*st* drawing, Captain Pascoe and Col. Charles R. Fox, under the watch of John H. Adams, an editor of *U.S. News and World Report*, set up the lottery.

They started out with 366 cylindrical capsules, one and a half inches long and one inch in diameter. The caps at the ends were round.

The men counted out 31 capsules and inserted in them slips of paper with the January dates. The January capsules were then placed in a large, square wooden box and pushed to one side with a cardboard divider, leaving part of the box empty.

The 29 February capsules were then poured into the empty portion of the box, counted again, and then scraped with the divider into the January capsules. Thus, according to Captain Pascoe, the January and February capsules were thoroughly mixed.

The same process was followed with each subsequent month, counting the capsules into the empty side of the box and then pushing them with the divider into the capsules of the previous months.

Thus, the January capsules were mixed with the other capsules 11 times, the February capsules 10 times and so on with the November capsules intermingled with others only twice and the December ones only once.

The box was then shut, and Colonel Fox shook it several times. He then carried it up three flights of stairs, a process that Captain Pascoe says further mixed the capsules.

The box was carried down the three flights shortly before the drawing began. In public view, the capsules were poured from the black box into the two-foot deep bowl.

Captain Pascoe said he did not know which end of the box he poured from. If he poured from the end where the capsules with the early months had been repeatedly shoved, these capsules might have fallen to the bottom of the bowl. Conversely, if he poured from the other end, the later months could have fallen to the bottom. This assumes that the shoving and shaking procedure did not adequately mix the capsules.

Once in the bowl, the capsules were not stirred. . . . The persons who drew the capsules last month generally picked ones from the top, although once in a while they would reach their hand to the middle or bottom of the bowl.

There is a serious question as to whether the capsules were adequately mixed. The description suggests that dates late in the year may have been drawn early, and dates early in the year may have been drawn late. In any case, we get the uneasy feeling that the mixing process was not thorough, but rather an inadequate mixing of "layers." By a *thorough* mixing, we mean that *every possible sequence of dates is equally likely to have been drawn*.

Since the number 366 of possible dates is quite large, it is difficult to visualize all the possible sequences (called **permutations** by statisticians). Let us take a much smaller case so that we can actually write down all possible permutations. For example, if only three numbers (say 1, 2, 3) are to be mixed so that all possible sequences or permutations are equally likely, then the list of permutations is easy to write down in an orderly fashion:

1	2	3
1	3	2
2	1	3
2	3	1
3	1	2
3	2	1

One simple systematic fashion of listing all permutations is to write the three-digit numbers in order of increasing size. For example, 123 is smaller than 132, which in turn is smaller than 213, . . . ; finally, the largest three-digit number 321 is listed.

A **fair** selection process means that each of the six permutations is equally likely to be selected. (Statisticians refer to such a selection process as a **random** selection process.) Since there are six permutations, each has a chance of one out of six (1/6) of being selected. Moreover, if it is necessary to make repeated selections, none of the selections depends on any one or more of the other selections, that is, all selections are **mutually independent**, as statisticians say.

On the other hand, if certain of the six possible permutations have a chance of occurring that differs from 1/6, then we say that the selection process is **biased** or **nonrandom**. For example, if the sequence 123 has a greater chance of being selected than 1/6, we say that the selection process tends to favor the outcome 123, and so the selection process is considered nonrandom or biased (in this case, in favor of the permutation 123).

It is very important to keep in mind that the fairness or, alternatively, the bias, refers to the selection process and not to the particular outcome that happens to turn up in any individual selection. This does *not* mean that we cannot take into account the particular outcomes that actually occur in applications of the selection process in deciding whether the selection process is fair or not.

Thus, returning to the draft lottery selection process, we suspect from the physical method of selecting the birthdates, that dates late in the year tend to be drawn early in the selection process, while dates early in the year tend to be drawn late in the selection process.

Do the outcomes displayed in Table 16.1 tend to support this suspicion? One way to answer this question is to compute the average lottery number for each month. For example, by adding the 31 numbers in the January column of Table 16.1 and then dividing by 31, we find that the average lottery number for January is

$$\frac{305 + 159 + 251 + \cdots + 349 + 164 + 211}{31} = \frac{6236}{31} = 201.2.$$

In a similar fashion, we obtain the average lottery number for February, March, . . . , December. The average lottery number for each of the 12 months is listed in Table 16.2.

By examining the entries in Table 16.2, we see a downward trend, especially from May to December; that is, the average lottery number tends to be high at the beginning of the year and low at the end. By drawing a plot of average lottery number against month (see Fig. 16.1), we see this downward trend even more clearly.

The downward trend in average lottery number by month displayed numerically in Table 16.2 and graphically in Fig. 16.1, along with the *New York Times* description of the mixing or randomizing process, yield strong evidence that the capsules were *not* sufficiently mixed to guarantee that the lottery drawing was fair. That is, the physical process of mixing did *not* ensure that all permutations of the 366 birthdates had an equal chance of occurring. Rather, early birthdates tended to receive high numbers, and late birthdates tended to receive low numbers.

Of course, statisticians do not reach *formal* conclusions simply by examining a table of numerical entries or viewing a graphical display. They rely on well-defined statistical procedures designed to yield exact probabili-

Table 16.1   The 1970 Selection Sequence by Month and Day

Day	Jan.	Feb.	Mar.	Apr.	May	June	July	Aug.	Sept.	Oct.	Nov.	Dec.
1	305	086	108	032	330	249	093	111	225	359	019	129
2	159	144	029	271	298	228	350	045	161	125	034	328
3	251	297	267	083	040	301	115	261	049	244	348	157
4	215	210	275	081	276	020	279	145	232	202	266	165
5	101	214	293	269	364	028	188	054	082	024	310	056
6	224	347	139	253	155	110	327	114	006	087	076	010
7	306	091	122	147	035	085	050	168	008	234	051	012
8	199	181	213	312	321	366	013	048	184	283	097	105
9	194	338	317	219	197	335	277	106	263	342	080	043
10	325	216	323	218	065	206	284	021	071	220	282	041
11	329	150	136	014	037	134	248	324	158	237	046	039
12	221	068	300	346	133	272	015	142	242	072	066	314
13	318	152	259	124	295	069	042	307	175	138	126	163
14	238	004	354	231	178	356	331	198	001	294	127	026

Day	Jan	Feb	Mar	Apr	May	Jun	Jul	Aug	Sep	Oct	Nov	Dec
15	017	089	169	273	130	180	322	102	113	171	131	320
16	121	212	166	148	055	274	120	044	207	254	107	096
17	235	189	033	260	112	073	098	154	255	288	143	304
18	140	292	332	090	278	341	190	141	246	005	146	128
19	058	025	200	336	075	104	227	311	177	241	203	240
20	280	302	239	345	183	360	187	344	063	192	185	135
21	186	363	334	062	250	060	027	291	204	243	156	070
22	337	290	265	316	326	247	153	339	160	117	009	053
23	118	057	256	252	319	109	172	116	119	201	182	162
24	059	236	258	002	031	358	023	036	195	196	230	095
25	052	179	343	351	361	137	067	286	149	176	132	084
26	092	365	170	340	357	022	303	245	018	007	309	173
27	355	205	268	074	296	064	289	352	233	264	047	078
28	077	299	223	262	308	222	088	167	257	094	281	123
29	349	285	362	191	226	353	270	061	151	229	099	016
30	164		217	208	103	209	287	333	315	038	174	003
31	211		030		313		193	011		079		100

Table 16.2 Average Lottery Number by Month

Month	Average Number
January	201.2
February	203.0
March	225.8
April	203.7
May	208.0
June	195.7
July	181.5
August	173.5
September	157.3
October	182.5
November	148.7
December	121.5

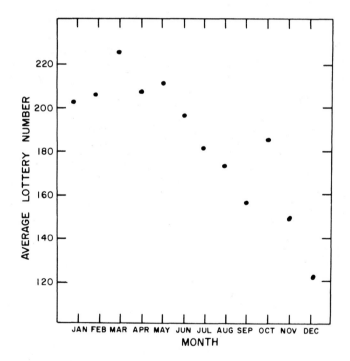

Figure 16.1 Average lottery number by month.

ties of various outcomes of interest. Unfortunately, we cannot describe in detail the exact statistical procedures that yield a specified assurance that a downward trend exists (or does not exist) in the average lottery number by month, since this would be beyond the scope of the book.

It does seem clear, however, that to arrive at a fair lottery of 366 birthdates (that is, a lottery in which all permutations of 1, 2, 3, . . . , 366 are equally likely) by physical mixing is a difficult task. Fortunately, tables of random numbers and random permutations generated by electronic computers can be used to achieve fair lotteries. [See Table A of Appendix I for random numbers and Moses and Oakford (1963) for random permutations.] Thus, the 1971 draft lottery was devised using such tables, with all possible sequences of birthdates being equally likely. The interested reader may obtain the details from the Rosenblatt and Filliben article in *Science* (1971).

Summary

1. There is evidence that the 1970 draft lottery was not fair. The method by which the physical mixing of the capsules was done suggests that dates late in the year tended to be drawn early in the selection process whereas dates early in the year tended to be drawn late in the selection process. A plot of the average lottery number by month seems to confirm the suggestion.

2. Fairness in selection processes can be achieved by careful planning which includes the use of tables of random numbers and/or tables of random permutations. The 1971 draft lottery was devised using such tables to ensure that all possible permutations of birthdays were equally likely to be selected.

Problems

1. What are other ways to investigate a trend using the Table 16.1 data?

2. Should February 29 be treated in the same way as are the other days of the year? After all, February 29 occurs only once every 4 years.

3. Census, life insurance, and other large collections of data show that more babies are born on certain dates than on other dates. How can this phenomenon be taken into account to achieve a draft lottery which is even more fair than the 1971 lottery. What is really meant in this context by "more fair"?

Table 16.3　The 1971 Selection Sequence by Month and Day

Day	Jan.	Feb.	Mar.	Apr.	May	June	July	Aug.	Sept.	Oct.	Nov.	Dec.
1	133	335	014	224	179	065	104	326	283	306	243	374
2	195	354	077	216	096	304	322	102	161	191	205	321
3	336	186	207	297	171	135	030	279	183	134	294	110
4	099	094	117	037	240	042	059	300	231	266	039	305
5	033	097	299	124	301	233	287	064	295	166	286	027
6	285	016	296	312	268	153	164	251	021	078	245	198
7	159	025	141	142	029	169	365	263	265	131	072	162
8	116	127	079	267	105	007	106	049	108	045	119	323
9	053	187	278	223	357	352	001	125	313	302	176	114
10	101	046	150	165	146	076	158	359	130	160	063	204
11	144	227	317	178	293	355	174	230	288	084	123	073
12	152	262	024	089	210	051	257	320	314	070	255	019
13	330	013	241	143	353	342	349	058	238	092	272	151
14	071	260	012	202	040	363	156	103	247	115	011	348

15	075	201	157	182	344	276	273	270	291	310	362	087
16	136	334	258	031	175	229	284	329	139	034	197	041
17	054	345	220	264	212	289	341	343	200	290	006	315
18	185	337	319	138	180	214	090	109	333	340	280	208
19	188	331	189	062	155	163	316	083	228	074	252	249
20	211	020	170	118	242	043	120	069	261	196	098	218
21	129	213	246	008	225	113	356	050	068	005	035	181
22	132	271	269	256	199	307	282	250	088	036	253	194
23	048	351	281	292	222	044	172	010	206	339	193	219
24	177	226	203	244	022	236	360	274	237	149	081	002
25	057	325	298	328	026	327	003	364	107	017	023	361
26	140	086	121	137	148	308	047	091	093	184	052	080
27	173	066	254	235	122	055	085	232	338	318	168	239
28	346	234	095	082	009	215	190	248	309	028	324	128
29	277		147	111	061	154	004	032	303	259	100	145
30	112		056	358	209	217	015	167	018	332	067	192
31	060		038		350		221	275		311		126

4. We listed in the text the $3 \times 2 \times 1 = 6$ permutations of the numbers 1, 2, 3. How many distinct permutations are there of the numbers:
   a. 1, 2, 3, 4?
   b. 1, 2, 3, 4, 5?
   c. 1, 2, 3, . . . , 366?

5.* Table 16.3 gives the results of the 1971 lottery. Read the article by Rosenblatt and Filliben (1971) for the details of the selection process, and then analyze the particular outcome of the process which is given in Table 16.3. Do you think the 1971 lottery was fair? Explain.

6. The mathematics faculty at a large university consists of 90 members. The department has the resources to send only 10 members to a special international meeting. How would you devise a process that makes all possible selections equally likely? Would you use a table of random numbers, a table of random permutations, or both? Explain.

7. What are some disadvantages of using the average lottery number computed for a specific month as indicative of the numbers assigned to the days of that month?

8.* How would you construct a table of random permutations?

9. Devise a selection process that tends to keep the average lottery numbers of the months approximately equal, but is not a fair selection process.

10. Suppose next year a draft lottery is to be conducted, carefully planned to make it fair. From what you know about draft lotteries, do you think some birthdates are more desirable than others? Explain.

References

1. S. E. Fienberg (1971). Randomization and social affairs: The 1970 draft lottery. *Science 171*, Jan. 22, 255–261.

2. L. E. Moses and R. V. Oakford (1963). *Tables of Random Permutations.* Stanford University Press, Stanford, Calif.

3. J. R. Rosenblatt and J. J. Filliben (1971). Randomization and the draft lottery. *Science 171*, Jan. 22, 306–308.

References 1 and 3 are interesting and for the most part nontechnical. Reference 2 is a nice book to own, but not to read.

Vignette 17
A Case of Bias?

The following is based on a statistical problem encountered by Joseph Gastwirth (1979). Identities have been changed to protect everybody who needs protection.

A female pharmacist at Lagranze Pharmaceuticals, Inc., filed a lawsuit against the company, complaining of sex discrimination. Her lawyer used Table 17.1 as evidence of discrimination. Table 17.1 is a list of pharmacists promoted to chief pharmacist between July 1965 and January 1973, with hire date, promotion date, and months from initial hire date to date of promotion.

To analyze the data in Table 17.1, we form a joint ranking of the times to promotion of the 24 males and the 2 females, the shortest time to promotion (5 months) getting rank 1, the second shortest time to promotion (7 months) getting rank 2, and so forth. The joint ranking looks like this:

Sex	M	M	M	M	M	M	M	M	M	M	M	M	M
Rank	1	2	3	4	5	6	7	8	9	10	11	12	13

Sex	M	M	M	M	M	M	M	M	M	M	F	F	M
Rank	14	15	16	17	18	19	20	21	22	23	24	25	26

The months to promotion for the 24 males can be viewed as a random sample of size 24 from the hypothetical **distribution** of times to promotion for male pharmacists at Lagranze. Similarly, the months to promotion for the two females can be viewed as a random sample of size 2 from the hypothetical distribution of times to promotion for female pharmacists. A charge of sexual discrimination against female pharmacists at Lagranze can be formulated in terms of the two distributions, with the contention that the times to promotion for females tend to be longer than the corresponding times for males, as reflected by the two distributions.

Statisticians call the assertion that the female and male distributions are the same, the **null hypothesis**. (The adjective "null" is used to indicate that

Table 17.1   Months to Promotion from Pharmacist to Chief Pharmacist

Name	Hire Date	Promotion Date	Months to Promotion
		Females	
J. L.	1/31	10/68	453
G. M.	6/49	7/68	229
		Males	
T. A.	12/68	11/72	47
E. B.	7/52	7/68	192
I. B.	8/70	10/71	14
J. B.	8/71	8/72	12
L. C.	1/70	3/71	14
E. C.	9/68	2/69	5
T. C.	12/65	1/69	37
K. D.	11/70	6/71	7
L. D.	1/62	10/67	69
M. D.	11/27	2/68	483
H. E.	10/68	8/71	34
A. G.	6/68	1/70	19
C. K.	9/70	10/72	25
E. L.	10/60	3/71	125
P. L.	9/67	7/70	34
J. M.	2/69	12/70	22
R. O.	12/69	1/72	25
A. P.	3/66	7/71	64
G. R.	3/70	5/71	14
R. R.	9/65	8/67	23
S. R.	10/68	7/70	21
C. S.	3/65	10/70	67
R. T.	5/65	4/69	47
C. W.	9/67	9/69	24

there is no difference.) The assertion that the female distribution tends to yield longer times than does the distribution for men is termed the **alternative hypothesis** of interest.

A **statistic** that can be used to decide whether the data support the null hypothesis or support the alternative hypothesis is the sum of the ranks obtained from the female times to promotion. It is clear that a very large value of this rank sum is more likely to occur when the alternative hypothesis is true. Furthermore, female ranks of 24, 25, with rank sum $24 + 25 = 49$, are (slightly) stronger evidence for the alternative hypothesis than are female ranks of 23, 25, with corresponding rank sum $23 + 25 = 48$.

Although a rank sum as large as 49 (such as we obtained from the data in Table 17.1) is more likely to occur when the alternative hypothesis is true than when the null hypothesis is true, it is possible to obtain by chance such a large rank sum even when the null hypothesis is true.

This leads to the following question: *What is the chance of obtaining a rank sum that is equal to or greater than 49 when females have the same distribution of time to promotion as do males?*

To answer this question, we use the fact that when females have the same distribution of length of time to promotion as do males, all possible joint rankings are equally likely to occur. We use the notation $\binom{26}{2}$ (read this as "26 choose 2") to denote the number of possible rankings when jointly ranking 2 females and 24 males.

The value $\binom{26}{2}$ can be obtained from the Table of Binomial Coefficients, Table B of Appendix I. It is equal to

$$\frac{26 \times 25}{2} = 325.$$

Next, note that there are only four possible joint rankings that can produce a sum as large as or larger than our observed sum of 49. The four are:

1. The ranking corresponding to the two females occupying positions 24 and 25 in the rank order.
2. The ranking corresponding to the two females occupying positions 23 and 26, also giving a rank sum of 49; $23 + 26 = 49$.
3. The ranking corresponding to the two females occupying positions 24 and 26, yielding a rank sum of $24 + 26 = 50$.
4. The ranking corresponding to the two females occupying positions 25 and 26, yielding a rank sum of $25 + 26 = 51$.

Since there are four joint rankings where the rank sum is 49 or greater, the chance of obtaining a rank sum that is greater than or equal to 49 when the null hypothesis is true is $4/325 = .0123$.

Thus, when the null hypothesis is true, there is only a very small chance (about 12 in 1,000) of observing a rank sum as large or larger than 49. Hence, we view the large rank sum of 49 as evidence supporting the alternative hypothesis that female waiting times do tend to be longer than their male counterparts.

Statisticians, who always hedge their bets, would state a conclusion along the following lines. Either the null hypothesis is true and we were unlucky enough to encounter a rare event or—what is more likely the case— the alternative hypothesis is true.

The foregoing analysis of the Table 17.1 data indicates that female pharmacists at Lagranze tend to wait longer for promotion to chief pharmacist than do their male counterparts. However, this does not establish beyond reasonable doubt that Lagranze discriminates against female employees. Other factors need to be considered, and other questions should be raised and answered. Among them are:

1. What are the promotion histories for men and women in other positions (manager, supervisor, buyer, and so on) within Lagranze?
2. What are the hiring records for Lagranze? How many female and male applicants are there per year, and what percentage of each group is hired?
3. Are men better trained than women when initially hired? Do they learn faster?

Summary

1. We wish to test whether two distributions are the same (null hypothesis) or, alternatively, one distribution lies to the right of the other (alternative hypothesis). A sample from each distribution is available, the two samples being independent. A simple test statistic that may be used is the rank sum statistic.
2. A significantly large value of the rank sum statistic leads to acceptance of the alternative hypothesis. Small and intermediate values lead to acceptance of the null hypothesis.
3. In some situations, a **two-sided alternative hypothesis** is appropriate. An example in the present context is as follows: The distribution of time until promotion is *not* the same for women as it is for men. The distribution of time until promotion for women may lie to the right of the corresponding distribution for men, or it may lie to the left. In the first case, women are being promoted at a slower rate; in the second case, at a faster rate (perhaps to avoid legal hassles or to attract potential female employees or to exploit this as a sales pitch in promoting cosmetic products).

The rank sum statistic would then be interpreted as follows: (1) A significantly large value would indicate women are being promoted more slowly. (2) A significantly small value would indicate women are being promoted more rapidly. (3) An intermediate value would indicate that women and men are being promoted at about the same rate. (Or, in the weasel-worded language of the perennially hedging statistician: "No significant evidence exists of a difference between the promotion-interval populations corresponding to men and women.")[†]

Problems

1. In Table 17.1, suppose that G. M.'s promotion date was 7/64 rather than 7/68. Perform a new analysis. In particular, what is the new sum of ranks resulting from the female times to promotion? If the null hypothesis is true, what is the chance of getting a rank sum as large or larger than this new sum?

2. Suppose that in addition to J. L. and G. M., Table 17.1 also contained a third female A. B. with hire date 7/50 and promotion date 7/68. Perform a new analysis based on the 3 females and the 24 males.

3. In our analysis we used the number of months to promotion. We did not make use of the information possibly present in the actual hire date or actual promotion date except for calculating the number of months to promotion. Explain how the actual hire date and promotion date might be useful in analyzing the company's treatment of female employees. (Hint: Consider, for example, a scenario where many females were promoted after long waits shortly after major legislation barring sexual discrimination was passed.)

4. In Table 17.1, suppose that J. L.'s hire date was 1/68 and G. M.'s hire date was 3/68. How would this affect your conclusions?

5. How would you, as the attorney for Lagranze, diminish the impact of Table 17.1?

6.* The defense attorney for Lagranze has argued that the times to promotion for J. L. and G. M. should *not* be considered a random sample of size 2 from the true times to promotion distribution for female pharmacists at Lagranze. The attorney says that these two values are not representative. The attorney claims that J. L. and G. M. did not do meritorious work and that more talented females would have been promoted sooner. What type of information would you need to investigate the validity of this claim?

7. We have used the sum of the female ranks as a measure of the degree to which the ranks achieved by the females tend to be small (or large). This sum is called the Wilcoxon rank sum statistic. [See chap. 4 of Hollander and Wolfe (1973) or chap. 13 of Brown and Hollander

---

[†] An experienced statistician's written statements bear an unpleasant similarity to printed leases issued by landlords. No matter what goes wrong, the burden is *not* on the preparer of the statement.

(1977).]† What other statistics would provide reasonable measures of the smallness (largeness) of the female ranks?

8. Do you envision some difficulties with the actual computation of the rank sum statistic when the sample sizes are large? Explain.

9. Suppose the sum of the ranks obtained from the female times to promotion were very small. What would this indicate?

10.* What is the expected value of the sum of the two female ranks if the null hypothesis is true?

References

1. B. W. Brown, Jr. and M. Hollander (1977). *Statistics: A Biomedical Introduction.* Wiley, New York.
2. J. Gastwirth (1979). Personal communication.
3. M. Hollander and D. A. Wolfe (1973). *Nonparametric Statistical Methods.* Wiley, New York.

Reference 1 provides an introduction to statistics that features examples from medicine and biology. Reference 3 describes statistical methods based on ranks.

---

†Problem within a problem: Suppose there are 100 texts which give good discussions of the Wilcoxon rank sum statistic. What is the probability that the references at this point would include a *non*-Hollander text? Answer: 0.   F. P.

Vignette 18
Do People Agree?

I don't know much about art, but I know what I like.

Overheard at the Museum of Modern Art

In many practical situations it is important to determine whether the members of a group agree or disagree on particular issues or preferences. For example:

1. A candidate for political office would benefit from knowing whether different ethnic groups in his district agree or disagree on the major issues of the day.
2. A distributor of wines would like to know if different sections of the country display differing preferences for the various wines.
3. The director of a program of recreational activities for senior citizens is interested in determining whether elderly women agree in their preferences as to the sex of the companions with whom they spend their leisure time.

Fortunately, for this last example, we can present in detail the data collected by Cindy Sutton, in the course of writing her 1976 Florida State University dissertation. We will also explain the main ideas used in analyzing these data statistically.

As part of her study of the preferences of elderly women, Sutton asked each of 14 white women in the 70-79 age group: Do you prefer to spend your leisure time with other women, with men, or in mixed company?

Notice that Sutton confined her question to members of a relatively homogeneous group. She was trying to avoid confusing the issue by mixing data obtained from groups that had differing preferences; for example, black elderly women or younger women might have different preference patterns.

Table 18.1   Leisure Time Companions Preferred by Elderly White Women

Elderly White Woman	Men	Women	Both Sexes
1	3	1	2
2	3	2	1
3	3	2	1
4	3	2	1
5	3	2	1
6	2	1	3
7	3	2	1
8	3	1	2
9	3	1	2
10	3	1	2
11	3	2	1
12	3	1	2
13	3	1	2
14	3	1	2
Sum of ratings	41	20	23

Each woman was asked to respond by writing 1 opposite her first (most preferred) choice, 2 opposite her second choice, and 3 opposite her third (least preferred) choice. The results are displayed in Table 18.1.

Thus, the first woman in the sample said her greatest preference was to spend her leisure time with other women, while her lowest preference was to spend her leisure time with men. Notice that six other women (#8, 9, 10, 12, 13, and 14) answered the question in precisely the same way; that is, exactly half the women interviewed exhibited the same preference pattern: 3 1 2. This degree of complete agreement in the sample suggests the following question: *Do elderly white women tend to agree in their preferences concerning the sex of their leisure time companions?*

In answering the question, we should keep aspects A, B, C, D below in mind.

A.   The data were obtained from a sample of 14 randomly selected elderly white women. The preference pattern in the population of elderly white women as a whole may differ somewhat from the preference pattern observed in the sample.

B.   Let us examine the actual sample outcome, keeping in mind the possible responses to Sutton's question. The six *possible* preference patterns are:

1	2	3
1	3	2
2	1	3
2	3	1
3	1	2
3	2	1

In the sample of 14 actual responses, 7 women gave the response 3 1 2, 6 gave the response 3 2 1, and 1 gave the response 2 1 3. The remaining three possible responses were not given by any of the women in the sample. That is, half the women agreed perfectly among themselves (the 7 women giving the response 3 1 2), whereas almost all of the other half (6 out of 7) agreed among themselves (the 6 women giving the response 3 2 1). Notice, also, that these two response patterns, 3 1 2 and 3 2 1, are rather similar; in both, men are preferred least as leisure time companions. The 14 women seem to show considerable agreement.

C.   To answer our question concerning agreement among the elderly white women, we should consider ways of measuring the degree of agreement. It turns out to be easier in some ways to think in terms of the degree of *disagreement*. To see this, suppose there were just six women in the sample, and each gave a different response as listed in the following table:

Woman	Response Patterns		
1	1	2	3
2	1	3	2
3	2	1	3
4	2	3	1
5	3	1	2
6	3	2	1
Total	12	12	12

It is clear that in this case the six women completely disagree. This is shown by the fact that every possible response pattern occurred exactly the same number of times (once, in this small sample).

More generally, if the sample size were larger, we would conclude that the women disagreed completely if each of the six different possible response patterns occurred about equally often.

Returning to Sutton's sample of 14 women, we see from Table 18.1 that the degree of disagreement was relatively small since the actual outcome was quite different from a one-sixth proportion for each of the 6 possible responses. This confirms our earlier conclusion that the disagreement shown in Sutton's sample was relatively small.

D. There is a less obvious, but more convenient, way to measure the degree of agreement among the women. You may have noticed that in the hypothetical sample of 6 women who disagreed completely, the total for each of the 3 categories (prefer men, prefer women, prefer both sexes) was the same, namely 12.

At the other extreme, suppose the six women in the hypothetical sample had shown complete agreement by responding as follows:

Woman	Response Patterns		
1	1	2	3
2	1	2	3
3	1	2	3
4	1	2	3
5	1	2	3
6	1	2	3
Total	6	12	18

Notice that in this case, representing complete agreement, the 3 totals are "spread out" as much as is possible (the first total, 6, is as small as is possible, the third total, 18, is as large as is possible, and the middle total is exactly half-way between the first and third totals). This illustrates a simple way of measuring the degree of agreement: *When the totals are spread out, the agreement is great; when the totals are close together, the agreement is small.*

Of course, the statistician has precise numerical measures of the degree of spread or, alternatively, of closeness of a set of observed totals. However, it would take us far afield to go into the details.

Going back to the data collected by Sutton, we see from Table 18.1 that the totals 41, 20, 23 display a good deal of spread. This suggests that the elderly white women agree quite well in their preferences.

Statisticians have developed precise statistical tests which take into account the size of the sample (14 in Sutton's case), the number of columns (three in Sutton's case), and the degree of spread among the column totals or, alternatively, the degree of spread among the numbers of individuals corresponding to the various possible response patterns (six possible response patterns in Sutton's case). From the test used, the statistician then concludes that the group is in agreement or not. In presenting the conclusion, the statistician also states the chance that it is erroneous. Actually there are two kinds of errors possible: (1) concluding that the elderly women agree when actually they disagree, and (2) concluding that the elderly women disagree when actually they agree. The possibility of reaching an incorrect conclusion is not a consequence of possible mistakes in calculation, but a consequence of the fact that the statistician is reaching a conclusion about a whole population based on a relatively small sample of individuals representing the population. This important feature of a statistical test, namely a precise knowledge of the chance of reaching either of the wrong conclusions (1) or (2), is not widely understood. Many people believe erroneously that because statisticians use precise statistical methods based on quantitative data, their conclusions are not subject to error.

Permutation data (such as the data of Table 18.1) arise in many contexts. We mentioned some examples at the beginning of this vignette. More examples and various approaches to analyzing such data are presented in an elegant manner by Diaconis (1985).

Summary

1. A frequently encountered problem is that of determining whether a group of people tend to agree. Examples include judges at a science fair, elderly women choosing their leisure activities, and tea tasters savoring various brands of tea.
2. Statisticians have devised precise measures, depending on the number of judges and the number of objects being judged, that enable one to reach a conclusion as to whether or not the judges are in agreement.
3. Two kinds of errors are possible: (1) concluding that the judges agree when they actually disagree, and (2) concluding that the judges disagree when they actually agree.

Problems

1. Table 18.2 is analogous to Table 18.1. It contains the responses of 13 elderly black women obtained by Sutton. Do these data indicate complete disagreement among the black women, or some element of agreement?
2.* Be creative and devise some ways to get a "ballpark" impression as to whether the elderly white women tend to agree with the elderly black women. [Hint: Since there are more white women (14) than black women (13), it is more meaningful to compare the averages, namely 41/14 with 30/13, 20/14 with 32/13, 23/14 with 16/13, rather than compare the totals (that is, 41 versus 30, 20 versus 32, 23 versus 16). To see the reasoning for preferring averages, consider for example, a case where there are 14 white women but 130 black women. Then the totals for the black women would be very different (larger) than the totals for the white women simply because there are more black women.]

Table 18.2 Leisure Time Companions Preferred by Elderly Black Women

Elderly Black Woman	Men	Women	Both Sexes
1	3	2	1
2	1	2	3
3	3	2	1
4	2	3	1
5	3	2	1
6	2	3	1
7	1	3	2
8	3	2	1
9	2	3	1
10	2	3	1
11	2	3	1
12	3	2	1
13	3	2	1
Column total	30	32	16

3. In 1977 Myles Hollander asked his 15 colleagues in the statistics department to rank the women playing detectives in the TV show Charlie's Angels, assigning score 1 for the favorite angel, and so on. The results for the 16 members of the department are given in Table 18.3. Do the statisticians seem to agree?

4. Table 18.4 contains the rankings of five brands of rum (A, B, C, D, E) by six Englishmen and seven Americans; 1 represents the most preferred, 5 the least.

   Do the English judges tend to agree among themselves? Do the American judges tend to agree among themselves? Do the English judges tend to agree with the American judges?

5.* Suppose, in addition to the English and American judges, a group of 10 Russian judges also ranked the brands of rum. How could we compare the three groups of judges?

6. How many preference patterns are there for the five brands of rum in Problem 4? More specifically, how many permutations of five distinct objects are there? (See Vignette 16 for a discussion of permutations).

7.* Explain why the notions of agreement and disagreement could have several reasonable interpretations.

8.* How does the situation discussed in this vignette differ from the situation where three judges are judging a diving competition where:
   a. Each of seven divers performs five dives?

Table 18.3 Preference Scores for the Women Playing Detectives in "Charlie's Angels"

Colleague	Kate Jackson	Cheryl Ladd	Jaclyn Smith
1	1	3	2
2	2	1	3
3	3	1	2
4	3	2	1
5	3	1	2
6	1	2	3
7	2	1	3
8	3	1	2
9	3	2	1
10	1	2	3
11	3	1	2
12	2	1	3
13	1	3	2
14	2	3	1
15	1	2	3
16	3	1	2

Table 18.4 Rankings of Rum Brands by Judges

	Rum				
	A	B	C	D	E
English Judge					
1	3	1	2	4	5
2	1	4	3	2	5
3	5	1	3	2	4
4	1	5	3	4	2
5	2	5	3	4	1
6	1	2	3	5	4
American Judge					
1	5	2	3	1	4
2	4	3	1	2	5
3	3	5	2	1	4
4	3	1	2	4	5
5	5	4	3	2	1
6	3	1	2	4	5
7	1	5	4	3	2

    b. Each judge independently awards a score of 1, 2, . . . , or 10 to each dive?

9. Return to Problem 4. What method would you suggest for picking the "best" brand of rum? Define "best."

10. Describe two different reasonable methods of selecting the winner at a science fair.

11.* Return to Problem 10. Suggest a method for selecting the winner when each judge cannot view each exhibit because of the large number of entries.

12.* The situation is the same as in Problem 11 (that is, many exhibits but too few judges). How could you obtain a complete ranking of the exhibits from best to worst?

References

1. P. Diaconis (1985). *Group Theory in Statistics.* Institute of Mathematical Statistics Lecture Notes-Monograph Series. (To appear.)
2. C. Sutton (1976). A study in relationships between selected background variables of older women and their expressed preferences for selected activity components. Ph.D. dissertation, Florida State University.

Reference 1 uses advanced mathematics to present beautiful treatments of fascinating statistical problems. Reference 2 is nonmathematical, and describes one example of a situation where questions of agreement arise.

Vignette 19

Do Overweight Men Tend to Have High Blood Pressure?

Many statistical studies are designed to explore the relationship between two random variables, and specifically to determine whether these variables are independent or dependent. Some examples are:

1. *Obesity and blood pressure:* Are obesity and blood pressure independent, or do overweight men tend to have high blood pressure?
2. *Tuna color and consumer preference:* Do consumers prefer light tuna over dark tuna? Alternatively, is their preference unrelated to the color?
3. *Infant crying and IQ:* Is the amount of crying in early infancy independent of intelligence, or do infants who cry a lot tend to have greater intelligence than do infants who cry less?
4. *Life length and environment:* Is the life length of a gyroscope on an aircraft independent of the temperature at which the system operates? Or does high temperature tend to shorten life length?

In this vignette we focus on the question of whether overweight men tend to have high blood pressure. To answer this question we use data collected by Farquhar and his associates (1974) in a random sample of Mexican-American males between the ages of 35 and 60 from the town of Watsonville, California (see Table 19.1). They are a portion of the data collected by Farquhar in the course of his studies of heart disease prevention. On each of the 44 men in the sample, two measurements were obtained. The degree of overweight was measured by the ratio of actual body weight to ideal body weight. The blood pressure recorded was systolic blood pressure (in mmHg). Note that neither variable is controlled or fixed in the study. The variables are simply measured as found among the men in the sample.

A basic tool for studying the relationship between two variables is a graph called the **scatterplot**. A scatterplot displays for each item the two

Table 19.1 Obesity and Blood Pressure in a Random Sample of 44 Mexican-American Males (aged 35 to 60) in a Small California Town

Obesity[a]	Blood Pressure[b]	Obesity[a]	Blood Pressure[b]
1.31	130	1.19	134
1.31	148	0.96	110
1.19	146	1.13	118
1.11	122	1.19	110
1.34	140	0.81	94
1.17	146	1.11	118
1.56	132	1.29	140
1.18	110	1.29	128
1.04	124	1.28	126
1.03	150	1.20	140
0.88	120	1.02	124
1.29	114	1.09	104
1.26	136	1.08	134
1.16	118	1.04	130
1.32	190	1.14	124
1.37	118	1.13	110
1.25	130	1.16	134
1.48	112	1.57	144
1.58	126	1.07	116
0.93	162	1.04	118
1.29	124	1.37	118
1.06	126	1.26	132

[a] Ratio of actual weight to ideal weight from New York Metropolitan Life Tables.
[b] Systolic blood pressure in mmHg.
Source: Reference 2.

variables measured. Figure 19.1 is a scatterplot for Farquhar's obesity-blood pressure data.

In Fig. 19.1 we started the horizontal scale at .80 and the vertical scale at 70 because they are convenient starting points for the obesity-blood pressure data. As in this example, generally the starting points are usually different for the two axes, and should be chosen for convenience and clear display of the data. To plot a point corresponding to a member of the sample, consider, for example, the fifteenth man. He has the obesity value 1.32 and the blood pressure value 190. We find on the horizontal axis the obesity value 1.32, and then we proceed vertically until we come to the height of 190 on the blood pressure scale. Putting a point at the coordinates (1.32, 190), we have plotted the two measurements for the fifteenth man.

Note some characteristics of the scatterplot.

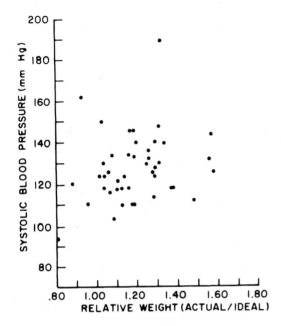

Figure 19.1   Scatterplot of blood pressure versus relative weight.

1. Each point represents one pair of observations. Thus, there are 44 points corresponding to the 44 men in the sample.
2. Each variable is represented by one of the axes. In the example, the horizontal axis represents obesity, and the vertical axis represents blood pressure.
3. The axes should be scaled so that (roughly) they are the same length; thus, the diagram looks square in shape. The horizontal axis should be scaled so that all values of the first variable observed in the sample can be included, and the vertical axis should be scaled so that all values of the second variable can be included. In this way, all of the items in the sample will appear in the scatterplot; that is, none will be off the graph.

In the obesity-blood pressure scatterplot, the dots tend to rise from the lower left to the upper right, although this tendency is slight rather than marked. An increase from the lower left toward the upper right indicates positive association between the two variables. That is, low values of blood pressure tend to accompany low values of obesity, and high values of blood pressure tend to accompany high values of obesity.

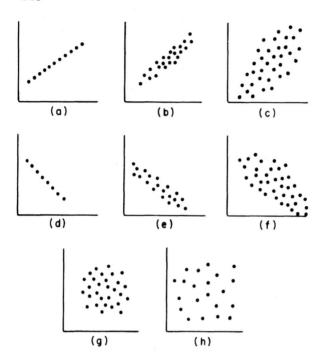

Figure 19.2   Scatterplots showing degree and type of association.

A drift from the upper left to the lower right in a scatterplot indicates a negative association between the two variables being plotted. That is, one variable tends to decrease as the other increases.

If a scatterplot displays a circular shape, or some other shape in which there is little or no trend, it indicates that there is very little, if any, dependence between the two variables.

Figure 19.2 consists of eight different scatterplots. The scatterplots in 19.2a, 19.2b, and 19.2c are indicative of situations in which the two variables plotted are positively associated. Figure 19.2a shows very strong positive association; Fig. 19.2b shows strong positive association, but not as strong as in 19.2a; and Fig. 19.2c shows modest positive association. On the other hand, Fig. 19.2d shows very strong negative association, Fig. 19.2e shows strong negative association, and Fig. 19.2f shows modest negative association. Figures 19.2g and 19.2h each indicate little or no relationship between the two variables plotted.

Statisticians have developed quantitative measures of the degree of relationship between two variables and its direction as exhibited by a sample as in Fig. 19.1. However, rather than go into details here, we restrict our discussion to the scatterplot, a very useful and visually interpretable tool.

A Warning About Causation. Although the scatterplot in Fig. 19.1 indicates that men who are overweight also tend to have high blood pressure, a study such as Farquhar's in which neither variable is manipulated or assigned cannot furnish definitive evidence concerning a causal relationship between the two variables. For the sample of Mexican-American males, the trend in the corresponding scatterplot could be explained in any of several ways. One could maintain that obesity causes high blood pressure, or that high blood pressure causes obesity, or that both are caused by a third factor, for example, heredity. If heredity were the cause, then changing either obesity or blood pressure might not change the other variable.

Summary

1. A scatterplot is a graph used as a basic tool for studying the relationship between two variables. A scatterplot displays for each item (person, object) in the sample the two variables measured.
2. If a scatterplot shows a drift from the lower left to the upper right, positive association between the two variables is indicated. That is, high values of the second variable tend to accompany high values of the first variable, and low values of the second variable tend to accompany low values of the first variable.
3. If a scatterplot shows a drift from the upper left to the lower right, negative association between the two variables is indicated. This means that one variable tends to decrease as the other increases.
4. If a scatterplot displays a circular shape, or some other shape in which there is little or no trend, this indicates that there is very little, if any, dependence between the two variables.
5. A study in which neither variable is manipulated or assigned cannot furnish definitive evidence concerning a causal relationship between the two variables.

Problems

1. Farquhar and his associates also were interested in the relationship between blood pressure and obesity in females. Table 19.2 presents the data for a random sample of 58 females. Graph the scatterplot. Does the scatterplot suggest dependence between obesity and blood pressure in females? How does this scatterplot compare with the corresponding one for males shown in Fig. 19.1?

Table 19.2 Obesity and Blood Pressure in a Random Sample of 58 Mexican-American Females Aged 35 to 60 in a Small California Town

Obesity[a]	Blood Pressure[b]	Obesity[a]	Blood Pressure[b]
1.50	140	1.54	130
1.59	150	1.43	128
1.43	130	1.74	128
1.63	132	1.51	118
2.39	150	1.36	110
1.50	112	1.37	148
0.92	138	1.67	162
1.17	116	1.03	128
1.33	124	1.32	108
1.09	112	1.56	116
1.24	116	1.33	104
1.44	110	1.56	122
1.23	160	1.25	98
1.50	140	1.24	110
1.34	124	1.27	118
2.04	138	1.57	116
1.13	118	1.30	118
1.11	104	1.32	138
1.38	114	1.41	142
1.35	138	1.21	124
1.42	170	1.20	120
1.55	144	1.15	118
1.33	108	1.43	122
1.22	108	1.24	112
1.07	98	1.28	126
0.97	112	1.75	138
1.26	100	2.20	136
1.65	120	1.64	136
1.01	118	1.73	208

[a] Ratio of actual weight to ideal weight from New York Metropolitan Life Tables.
[b] Systolic blood pressure in mmHg.
Source: Reference 2.

2. Karelitz, Fisichelli, Costa, Karelitz, and Rosenfeld (1964) studied the relation between crying activity in early infancy and speech and intellectual development at age 3 years. [Our knowledge of the Karelitz *et al.* paper comes from reading Lehmann (1975).] A visual tally of cry counts was taken for a sample of 38 infants aged 4–7 days. Three years later the Stanford-Binet IQ scores of these 38 children were obtained.

Table 19.3   Cry Counts and Stanford-Binet IQ Scores

Subject	Sex	Cry Count	Stanford-Binet IQ Score
1	F	10	87
2	M	20	90
3	M	17	94
4	F	12	94
5	M	12	97
6	F	16	100
7	F	19	103
8	M	12	103
9	M	9	103
10	M	23	103
11	M	13	104
12	F	14	106
13	M	16	106
14	M	27	108
15	M	18	109
16	F	10	109
17	M	18	109
18	F	15	112
19	M	18	112
20	M	23	113
21	F	15	114
22	M	21	114
23	M	16	118
24	F	9	119
25	F	12	119
26	M	12	120
27	F	19	120
28	F	16	124
29	F	20	132
30	F	15	133
31	F	22	135
32	F	31	135
33	F	16	136
34	M	17	141
35	M	30	155
36	F	22	157
37	F	33	159
38	F	13	162

Source: Reference 4.

The data are presented in Table 19.3. For convenience, the children are listed in increasing order of their IQs.

Draw a scatterplot for the males and one for the females. Do they roughly agree? Do they suggest a relationship between the amount of crying in early infancy and IQ scores at age 3?

3. Blot and Fraumeni (1976) point out that there is striking geographical variation in lung cancer mortality within the United States, with many of the highest rates among men in southern coastal areas, particularly in Louisiana, rather than in the urban, industrialized northeast. Table 19.4 lists lung cancer mortality rates among white males and females

Table 19.4 Lung Cancer Mortality and Degree of Urbanization in Louisiana— Lung Cancer Mortality Rate, 1950–1969 (Deaths/Year/100,000 Population)

Parish	Males	Females	Percentage Urban
Acadia	56.98	11.66	55.8
Allen	48.37	5.62	33.3
Ascension	55.39	4.04	33.3
Assumption	52.89	4.83	0
Avoyelles	47.72	4.64	25.1
Beauregard	36.96	9.87	37.5
Bienville	35.35	2.72	15.1
Bossier	48.14	6.86	66.0
Caddo	53.92	7.57	80.8
Calcasieu	54.10	8.24	73.8
Caldwell	34.62	8.94	0
Cameron	36.41	9.11	0
Catahoula	59.30	5.84	0
Claiborne	42.75	2.88	39.6
Concordia	66.85	5.04	43.3
De Soto	42.54	9.44	24.1
East Baton Rouge	44.43	6.26	85.1
East Carroll	51.72	8.21	40.1
East Feliciana	20.80	1.34	0
Evangeline	60.59	14.87	33.0
Franklin	38.84	7.09	17.0
Grant	40.65	10.51	0
Iberia	62.48	4.87	67.0
Iberville	46.02	5.73	25.6
Jackson	39.99	5.63	24.3
Jefferson	62.52	7.91	94.1

Table 19.4 *(Continued)*

Parish	Males	Females	Percentage Urban
Jefferson Davis	52.60	10.10	62.8
Lafayette	51.60	9.52	55.6
Lafourche	58.00	5.45	41.5
La Salle	48.92	3.31	0
Lincoln	28.71	6.77	60.0
Livingston	52.11	5.72	22.1
Madison	54.41	3.34	57.1
Morehouse	38.64	3.02	45.1
Natchitoches	32.36	6.96	39.1
Orleans	64.10	7.73	100.0
Ouachita	49.01	5.33	79.1
Plaquemines	50.12	4.61	34.5
Pointe Coupee	38.62	6.46	17.6
Rapides	41.09	5.33	52.6
Red River	36.79	5.50	0
Richland	56.13	6.60	27.6
Sabine	23.34	4.01	17.0
St. Bernard	71.80	5.44	66.0
St. Charles	60.16	4.36	22.1
St. Helena	34.63	4.60	0
St. James	41.97	5.98	17.8
St. John Baptist	36.06	2.60	47.8
St. Landry	52.77	7.35	35.3
St. Martin	61.03	8.42	33.6
St. Mary	56.38	5.07	59.3
St. Tammany	46.03	8.99	33.8
Tangipahoa	54.07	7.06	35.6
Tensas	68.52	2.43	0
Terrebonne	68.55	3.53	52.1
Union	29.47	5.79	15.5
Vermilion	44.92	5.70	40.3
Vernon	33.72	6.11	25.6
Washington	48.40	6.89	55.8
Webster	39.86	4.84	48.3
West Baton Rouge	54.27	5.71	39.1
West Carroll	36.08	6.24	0
West Feliciana	34.52	16.45	0
Winn	33.06	7.12	43.8

Sources: References 6 and 7.

over the 20-year period 1950–1969 for Louisiana's 64 parishes, along with the percentage of the population in each parish living in urban areas in 1960. Graph the scatterplot of male mortality rate versus percentage urban. Does the scatterplot indicate that male mortality rates increase with degree of urbanization? Graph the scatterplot of female mortality rate versus percentage urban. Does this scatterplot indicate that female mortality rates increase with degree of urbanization?

4. Fraumeni (1968) considered the possibility of a relationship between cigarette smoking and certain types of cancer. The data in Table 19.5 are based in part on 1960 records from the 43 states and the District of Columbia, all of which charged a cigarette tax.

Table 19.5   Cigarette Smoking and Cancer Death Rates

State	Cigarettes Per Person[a]	Deaths Per Year Per 100,000 Population			
		Bladder Cancer	Lung Cancer	Kidney Cancer	Leukemia
Alabama	1,820	2.90	17.05	1.59	6.15
Arizona	2,582	3.52	19.80	2.75	6.61
Arkansas	1,824	2.99	15.98	2.02	6.94
California	2,860	4.46	22.07	2.66	7.06
Connecticut	3,110	5.11	22.83	3.35	7.20
Delaware	3,360	4.78	24.55	3.36	6.45
District of Columbia	4,046	5.60	27.27	3.13	7.08
Florida	2,827	4.46	23.57	2.41	6.07
Idaho	2,010	3.08	13.58	2.46	6.62
Illinois	2,791	4.75	22.80	2.95	7.27
Indiana	2,618	4.09	20.30	2.81	7.00
Iowa	2,212	4.23	16.59	2.90	7.69
Kansas	2,184	2.91	16.84	2.88	7.42
Kentucky	2,344	2.86	17.71	2.13	6.41
Louisiana	2,158	4.65	25.45	2.30	6.71
Maine	2,892	4.79	20.94	3.22	6.24
Maryland	2,591	5.21	26.48	2.85	6.81
Massachusetts	2,692	4.69	22.04	3.03	6.89
Michigan	2,496	5.27	22.72	2.97	6.91
Minnesota	2,206	3.72	14.20	3.54	8.28
Mississippi	1,608	3.06	15.60	1.77	6.08
Missouri	2,756	4.04	20.98	2.55	6.82
Montana	2,375	3.95	19.50	3.43	6.90
Nebraska	2,332	3.72	16.70	2.92	7.80
Nevada	4,240	6.54	23.03	2.85	6.67

Table 19.5   *(Continued)*

| State | Cigarettes Per Person[a] | Deaths Per Year Per 100,000 Population | | | |
		Bladder Cancer	Lung Cancer	Kidney Cancer	Leukemia
New Jersey	2,864	5.98	25.95	3.12	7.12
New Mexico	2,116	2.90	14.59	2.52	5.95
New York	2,914	5.30	25.02	3.10	7.23
North Dakota	1,996	2.89	12.12	3.62	6.99
Ohio	2,638	4.47	21.89	2.95	7.38
Oklahoma	2,344	2.93	19.45	2.45	7.46
Pennsylvania	2,378	4.89	22.11	2.75	6.83
Rhode Island	2,918	4.99	23.68	2.84	6.35
South Carolina	1,806	3.25	17.45	2.05	5.82
South Dakota	2,094	3.64	14.11	3.11	8.15
Tennessee	2,008	2.94	17.60	2.18	6.59
Texas	2,257	3.21	20.74	2.69	7.02
Utah	1,400	3.31	12.01	2.20	6.71
Vermont	2,589	4.63	21.22	3.17	6.56
Washington	2,117	4.04	20.34	2.78	7.48
West Virginia	2,125	3.14	20.55	2.34	6.73
Wisconsin	2,286	4.78	15.53	3.28	7.38
Wyoming	2,804	3.20	15.92	2.66	5.78
Alaska	3,034	3.46	25.88	4.32	4.90

[a] Estimated from cigarette tax revenues.
Source: Reference 3.

Draw four scatterplots: cigarette consumption versus death rate for bladder cancer, cigarette consumption versus death rate for lung cancer, cigarette consumption versus death rate for cancer of the kidney, and cigarette consumption versus death rate for leukemia. Do the states with high cigarette consumption tend to have high death rates from lung cancer? What relationships between cigarette consumption and the other types of cancer are suggested?

5. Korsakoff's syndrome is an organic brain disease characterized by varying degrees of amnesia. McEntee and Mair (1978) have studied the relationship between the cerebrospinal fluid (CSF) concentration of certain brain metabolites and the severity of memory impairment in nine patients having Korsakoff's syndrome. Their measure of memory impairment is IQ minus MQ, where IQ is the full-scale intelligence quotient derived from the Wechsler Adult Intelligence Scale and MQ is the memory quotient derived from the Wechsler Memory Scale.

Roughly speaking, the larger the value of IQ minus MQ, the more memory impairment is considered to be present. Table 19.6 contains values of IQ, MQ, and the concentrations (in nanograms per milliliter) of certain brain metabolites: MHPG, VMA, HVA, and 5-HIAA.

Compute IQ minus MQ for each patient. Draw a scatterplot of "IQ minus MQ" versus spinal fluid MHPG. Does the scatterplot indicate negative association, no association, or positive association between memory impairment (as measured by IQ minus MQ) and the concentration of MHPG? Continue your data exploration by drawing three more scatterplots:

a. IQ minus MQ versus VMA
b. IQ minus MQ versus HVA
c. IQ minus MQ versus 5-HIAA

McEntee and Mair (1978) point out that lesions associated with Korsakoff's syndrome are often located along the pathways of neurons containing these metabolites. McEntee and Mair find that the data of Table 19.6 suggest that the memory disorder of Korsakoff's syndrome may be associated with damage to these pathways. See their paper for more details; somehow, we cannot recall them at this moment.

6. In studying the relationship between two variables, we use two basic methods. Under the correlation method, neither variable is held fixed by the experimenter. Under the regression method, deliberately chosen values of one variable are fixed, and the corresponding values of the

Table 19.6   Full-Scale Intelligence Quotient (IQ), Memory Quotient (MQ), and Concentration of Certain Brain Metabolites for Nine Korsakoff's Syndrome Patients

Patient	IQ	MQ	Metabolite			
			MHPG	VMA	HVA	5-HIAA
1	89	61	3.71	0.76	26	17.1
2	104	80	8.66	0.47	40	17.0
3	106	80	6.47	1.19	48	20.9
4	127	88	5.36	0.56	75	65.2
5	122	102	9.75	0.84	25	32.7
6	106	79	5.93	0.22	31	10.6
7	87	64	6.95	0.71	25	17.0
8	89	60	2.96	0.43	21	17.0
9	90	59	5.16	0.43	23	14.9

Source: Reference 8.

other variable are measured. Describe two correlation problems and two regression problems.

7.* Review preceding vignettes and peek ahead at the subsequent vignettes. Do you find any problems that could be cast in the framework of:
   a. A correlation problem?
   b. A regression problem?

8. Suppose one or two men were added to Farquhar's sample. Do you think their measurements would significantly affect the scatterplot and the interpretations reached from the scatterplot? Explain.

9. How would you interpret a scatterplot that looks like the letter "V".

10. How would you interpret a scatterplot that looks like Λ?

References

1. W. J. Blot and J. F. Fraumeni, Jr. (1976). Geographic patterns of lung cancer: Industrial correlations. *American Journal of Epidemiology 103*, 539–550.

2. J. W. Farquhar (1974). Personal communication.

3. J. F. Fraumeni, Jr. (1968). Cigarette smoking and cancers of the urinary tract: Geographic variation in the United States. *Journal of the National Cancer Institute 41*, 1205–1211.

4. S. Karelitz, V. R. Fisichelli, J. Costa, R. Karelitz, and L. Rosenfeld (1964). Relation of crying in early infancy to speech and intellectual development at age three years. *Child Development 35*, 769–777.

5. E. L. Lehmann (1975). *Nonparametrics: Statistical Methods Based on Ranks.* Holden-Day, San Francisco.

6. T. J. Mason and F. W. McKay (1974). U.S. Cancer Mortality by County: 1950–69. U.S. Government Printing Office, Washington, D.C.

7. U.S. Bureau of Census (1963). U.S. Census of Population: 1960. U.S. Government Printing Office, Washington, D.C.

8. W. J. McEntee and R. G. Mair (1978). Memory impairment in Korsakoff's psychosis: A correlation with brain noradrenergic activity. *Science 202* Nov. 24, 905–907.

References 1–8 contain the data sets used in this vignette. Reference 5 is an excellent intermediate-level book which treats statistical methods based on ranks.

Vignette 20

Planning Experiments

"My husband's a professor at M.I.T."
"Really? What's his field?"
"Statistics."
"Oh, a real brain. I'm always self-conscious when I meet that sort of mind. I can barely add a column of figures."
"Neither can Bob." Sheila smiled. "That's my job at the end of every month."

Erich Segal, *Man, Woman and Child*

Many people believe that the chief function of the statistician is to analyze data *after* they have been collected by the scientist, engineer, biomedical researcher, and so forth. But actually, the statistician makes a very important contribution at an earlier stage. He helps the researcher in planning or **designing the experiment** which will yield the data to be analyzed. Using a properly designed experiment, the researcher gathers information of much higher quality than otherwise. We can explain this most easily by examples.

Suppose a biomedical researcher is studying the effect (if any) of smoking on lung cancer. An obvious method of collecting data to study the relationship is to determine from public health records the proportion of deaths from lung cancer among smokers and the proportion of deaths from lung cancer among nonsmokers. If smokers experienced a significantly higher proportion of lung cancer deaths, it would seem to indicate that smoking and lung cancer are positively linked.

A number of difficulties are present in performing such a **retrospective** study. (A retrospective collection of data is a set of data obtained historically, as contrasted with a set of data obtained from a planned experiment.)

158

Virginia Smoliar

"Professor Weiss—kuff, kuff—the relationship between—hack, kuff—cigarette smoking—hack, kuff—and lung cancer is purely—kuff—statistical, not—hack, causal."

1. The records may not specify for each individual whether he is a smoker or not. Similarly, the cause of death may not always be specified. That is, the data may be **incomplete**.
2. There may be a much larger number of smokers than nonsmokers, or vice versa. This would tend to make one proportion a more reliable estimate than the other.
3. There may be errors in the data. For example, a death may be attributed to the wrong cause.
4. The data may not be as refined as desired. For example, the researcher may wish to know whether a smoker used cigarettes, cigars, or a pipe, and the extent of use. However, the data may not contain such details.
5. The two categories, smokers and nonsmokers, might differ in ways which affect tendencies to contract lung cancer. For example,

certain religions forbid smoking; perhaps the life style of the members of these religions differs in ways which have some effect on the lung cancer rate. Thus, the cause of the higher rate of lung cancer among smokers might be the result of factors other than smoking.

It seems clear that using retrospective data to reach scientific conclusions may be less than ideal. What is the alternative?

An alternative approach leading to conclusions having much greater precision is to obtain the data from a **planned experiment**. Let us illustrate this idea. Suppose the researcher matches pairs of individuals so that (1) one individual of the pair is a smoker, (2) the other individual of the pair is a nonsmoker, and (3) the two individuals *are as alike as possible in all other ways*. For example, the two individuals might be of the same age, of the same sex, of the same nationality, and have similar occupations. Of course, the researcher tries to obtain as many **matched pairs** as possible. He then determines the rate of lung cancer among the smokers and compares it with the rate of lung cancer among the nonsmokers.

It is clear that the effect of extraneous factors is very largely reduced. Since the two individuals being compared within each pair are as alike as possible, the effect of smoking is clearly brought out. The conclusion reached from the data is much more reliable.

There are two ways to use the method of matched pairs. One way is to form the matched pairs from the historical data available. An alternative way is to form matched pairs of individuals currently available to the experimenter and to observe in the future the proportion of smokers dying of lung cancer and the proportion of nonsmokers dying of lung cancer.

The advantage of the second method is that matching can be carried out much more carefully and thoroughly. The experimenter selects the two individuals within each pair so that they are alike with regard to as many criteria as he deems significant. Under the first method (relying on the historical data), the statistician is much more limited in his ability to achieve good matching within pairs since he is at the mercy of the historical data. That is, if necessary information is unavailable, he is generally unable to obtain it.

Let us consider another aspect of planned experimentation using matched pairs. We illustrate with an example from a somewhat different area of medical research. Suppose a medical researcher wishes to determine whether a new drug is more effective in the treatment of bone cancer than is the standard drug. He very carefully selects 25 pairs of patients such that the two individuals in each pair are as much alike as possible. For each matched pair, the researcher must decide which patient gets the new treatment and which patient gets the standard treatment.

This is a delicate stage in the planning of the experiment. There are a number of pitfalls to be avoided, some obvious, some not so obvious.

1. If the medical researcher *subjectively* chooses the patient to receive the new drug which he has developed, he may unconsciously choose the patient who seems to be in better shape. That is, without being aware of his bias, in his zeal to demonstrate that his new drug is superior, he may try to give it the benefit of using it on the better patient of the pair. (It is true that the two patients are as alike as possible, but still, common sense tells us that they cannot be exactly alike.)

2. If the patient learns that he is being treated with a new drug rather than with the standard drug, this information may affect his recovery. Depending on his faith (or lack of faith) in his doctor, he may become more optimistic (or more pessimistic) about recovery. This emotional attitude may play a role in his medical progress.

3. Even the attitude of the nurses may play a role in the experiment. Their knowledge of which patient is receiving the new drug may interfere with their objectivity in administering the drug or reporting their findings in their examinations of the patients.

To avoid these sources of bias, the researcher performs a **double-blind**[+] **experiment**. First of all, he picks the patient in each matched pair to receive the new drug using a table of random numbers. (For example, he decides in advance that the patient whose name comes first alphabetically will receive the new drug if the randomly selected number is odd; otherwise the other patient in the matched pair will receive the new drug.) This avoids bias on the part of the experimenter. A second step taken to avoid bias is to arrange the experiment so that the patients, nurses, and even the medical researcher do not know which patients are receiving the new drug and which patients are receiving the standard drug until after the experiment and the individual evaluations have been completed. This requires some extra trouble and careful security, but it can be done.

The double-blind experiment using matched pairs of patients is an effective method of comparing two drugs, two treatments, or two categories in general. By the design of the experiment, bias is avoided and the precision of the comparison is increased.

We have illustrated the notion of deliberately designing an experiment to obtain the data needed for reaching reliable conclusions, using one simple type of experiment as an example. Actually, there is a great variety of

[+]This is not an experiment in which the "blind lead the blind" or an experiment in which nobody knows what the hell is going on.

experimental designs that have been evolved by statisticians to be used for obtaining information in a precise, unbiased fashion. Accompanying each experimental design is a corresponding method of statistical analysis which not only yields conclusions, but also yields an estimate of the precision of each conclusion.

Summary

Statisticians not only analyze data after they have been collected, but also are very helpful in planning and designing the experiments that will yield the data. We have described a particular type of experimental design in which matched pairs of specimens are used. The two members of the pair are chosen as alike as possible; one member receives one type of treatment, the other member receives the other type of treatment. Thus, in comparing the two treatments, extraneous factors are eliminated as much as possible.

Additional improvement is possible by the use of the double-blind method. The experiment is set up so that during the course of the experiment, none of the subjects or even the experimenter knows which member of the pair is receiving the first type of treatment and which member is receiving the second type of treatment. Of course, after all the data have been collected, the proper identifications are made linking subject and treatment. The purpose of the "blindness" during the experiment is to eliminate even the most subtle sources of bias.

Many other types of experimental designs have been developed by statisticians.

Problems

1. A cancer researcher wishes to determine whether a certain food preservative is a carcinogen (cancer-producing agent). He selects 50 pairs of mice, each pair coming from a common litter. Using a table of random numbers he determines which mouse of the pair will be injected with the preservative and which mouse will not. After the injections have been performed, he examines the mice at regular intervals to see whether tumors have formed and whether they are malignant or not.

   Compare this procedure with the following procedure used by a colleague in a similar experiment: 100 mice are selected and divided into 2 groups of 50 each. One group receives an injection, the other does not. As before, the experimenter examines the two groups at regular intervals and records his findings.

   a. Which procedure is more likely to isolate the effect of the food preservative? Explain your answer.

b. Can you suggest any improvements in the procedure followed by the first experimenter?

c. A public health scientist has noticed that the food preservative is used heavily in a tonic popular with aged people. He compares the cancer rate among users of the tonic with the cancer rate among nonusers, discovering that the incidence of cancer is a good deal higher among users of the tonic. From this evidence, he concludes that the food preservative is a carcinogen.

What pitfall exists in the reasoning of the public health scientist?

Assuming the data are sufficiently detailed, what modifications in the data analysis would lead to a more sound conclusion? Hint: How can the scientist take into account the effect of age on the occurrence of cancer?

2. A medical researcher wishes to determine whether a certain diet plan is effective in reducing weight. Twenty-five overweight subjects have volunteered to try out the new diet. He weighs each subject before the subject begins the new diet and again 6 weeks later, during which time the subject has been using the new diet. He then reaches his conclusions from the data.

Express your opinion on the following possible additions or changes in the study:

a. The before and after weighings occur at the same time of day.

b. The subject is weighed once each week at the same time of day.

c. The weighing is performed by a nurse.

d. Precautions are taken to ensure that the subjects do not "supplement" the diet or deviate from it in any way.

3. Can you think of some situations in which the double-blind approach is extremely difficult (if not impossible) to implement? Explain.

4. Is a matched-pairs design feasible for answering the questions posed in Vignette 17, "A Case of Bias?"? Explain.

5. Many researchers prefer to use twins in matched-pairs experiments. Explain.

6. The round-robin tournament (see Vignette 9) is an experiment that requires careful planning. Explain.

7. In this vignette we have mentioned some difficulties associated with retrospective studies. What are some advantages of retrospective studies?

8.* A medical researcher has hypothesized that operating room nurses tend to have more spontaneous abortions than general duty nurses and that this is due to anesthetic gases escaping into the operating room atmosphere. How would you plan a study to investigate this hypothesis? Would you favor a retrospective or prospective study? Explain.

9.* Sketch a study to investigate whether high levels of aspirin intake lead to an increased incidence of stomach cancer.

10.* Describe some experiments in which it is natural and judicious for the same patient to receive the standard treatment and the new treatment. (That is, where a "matched pair" really consists of one individual who receives both treatments.) What assumptions should be satisfied in such situations?

11. Sketch a study to determine whether "Sunglow" promotes tanning better than "Bronzetone."

12. Sketch a study to determine whether 20 minutes of batting practice per day is better than 10 minutes of batting practice per day.[†]

[†] This is carrying the cautious statistical approach just a bit too far! Do you need to assemble data to conclude what is obvious a priori? F. P.

It's not as obvious as you make it out to be. A recent study was conducted on the effect of practice on water polo proficiency. One team practiced underwater steadily for 5 minutes, while a second team practiced under water steadily for 2½ minutes. The first team unfortunately all succumbed, while the second team survived, though in a condition resembling dank vegetables. Modern science is based on observation, not simply on rational judgment. M. H.

ESTIMATING PROBABILITIES

"You got a ninety percent chance," he said.
Osano said quickly, "How do you get that figure?" He always did that whenever somebody pulled a statistic on him. He hated statistics.

Mario Puzo, *Fools Die*

Copyright, 1979, *American Scientist*.

Vignette 21
Life Tables

Copyright, 1980, United Features Syndicate, Inc.

"I'm tired of Bernice picking the color of every new car we buy."

"But Gordy, she also pays for the cars."

"Yes, but you know father wanted me to have that money and now we'll have to wait till twilight time. She's 73 and playing golf, her mother is 92, and her grandmother hit triple digits before calling it a century. Longevity is killing us."

Joyce McShane, *California Curry*

Table 21.1 is a **life table** for the population of the United States corresponding to the 1959–1961 era. What kind of information does it contain? How can this information be used?

First, let us make sure we understand what the entries mean. Then we will discuss what uses can be made of the table.

The entry in the first column represents the age in yearly intervals. For example, if you want information concerning youths between their twentieth and twenty-first birthdays, you would look at the row corresponding to the age interval 20–21. The entry (.00115) in the second column is quite well described by the heading: it represents the proportion of those individuals who have reached their twentieth birthday who, on the average, will die

Table 21.1 Complete Life Table for the Total Population of the United States: 1959–1961

Age Interval (Years)[a]	Proportion Dying: Proportion of Persons Alive at Beginning of Age Interval Dying During Interval	Of 100,000 Born Alive		Average Remaining Lifetime: Average Number of Years of Life Remaining at Beginning of Age Interval
		Number Alive at Beginning of Age Interval	Number Dying During Age Interval	
0–1	.02593	100,000	2,593	69.89
1–2	.00170	97,407	165	70.75
2–3	.00104	97,242	101	69.87
3–4	.00080	97,141	78	68.94
4–5	.00067	97,063	65	67.99
5–6	.00059	96,998	57	67.04
6–7	.00052	96,941	50	66.08
7–8	.00047	96,891	46	65.11
8–9	.00043	96,845	42	64.14
9–10	.00039	96,803	38	63.17
10–11	.00037	96,765	36	62.19
11–12	.00037	96,729	36	61.22
12–13	.00040	96,693	39	60.24
13–14	.00048	96,654	46	59.26
14–15	.00059	96,608	57	58.29
15–16	.00071	96,551	68	57.33
16–17	.00082	96,483	80	56.37
17–18	.00093	96,403	89	55.41
18–19	.00102	96,314	98	54.46
19–20	.00108	96,216	105	53.52
20–21	.00115	96,111	110	52.58
21–22	.00122	96,001	118	51.64
22–23	.00127	95,883	122	50.70
23–24	.00128	95,761	123	49.76
24–25	.00127	95,638	121	48.83
25–26	.00126	95,517	120	47.89
26–27	.00125	95,397	120	46.95
27–28	.00126	95,277	120	46.00
28–29	.00130	95,157	123	45.06
29–30	.00136	95,034	129	44.12
30–31	.00143	94,905	136	43.18
31–32	.00151	94,769	143	42.24
32–33	.00160	94,626	151	41.30
33–34	.00170	94,475	160	40.37

Table 21.1   *(Continued)*

Age Interval (Years)[a]	Proportion Dying: Proportion of Persons Alive at Beginning of Age Interval Dying During Interval	Of 100,000 Born Alive		Average Remaining Lifetime: Average Number of Years of Life Remaining at Beginning of Age Interval
		Number Alive at Beginning of Age Interval	Number Dying During Age Interval	
34–35	.00181	94,315	171	39.44
35–36	.00194	94,144	183	38.51
36–37	.00209	93,961	196	37.58
37–38	.00228	93,765	214	36.66
38–39	.00249	93,551	232	35.74
39–40	.00273	93,319	255	34.83
40–41	.00300	93,064	279	33.92
41–42	.00330	92,785	306	33.02
42–43	.00362	92,479	335	32.13
43–44	.00397	92,144	366	31.25
44–45	.00435	91,778	400	30.37
45–46	.00476	91,378	435	29.50
46–47	.00521	90,943	473	28.64
47–48	.00573	90,470	519	27.79
48–49	.00633	89,951	569	26.94
49–50	.00700	89,382	626	26.11
50–51	.00774	88,756	687	25.29
51–52	.00852	88,069	751	24.49
52–53	.00929	87,318	811	23.69
53–54	.01005	86,507	870	22.91
54–55	.01082	85,637	926	22.14
55–56	.01161	84,711	983	21.37
56–57	.01249	83,728	1,047	20.62
57–58	.01352	82,681	1,117	19.87
58–59	.01473	81,564	1,202	19.14
59–60	.01611	80,362	1,295	18.42
60–61	.01761	79,067	1,392	17.71
61–62	.01917	77,675	1,489	17.02
62–63	.02082	76,186	1,586	16.34
63–64	.02252	74,600	1,680	15.68
64–65	.02431	72,920	1,773	15.03
65–66	.02622	71,147	1,866	14.39
66–67	.02828	69,281	1,959	13.76
67–68	.03053	67,322	2,055	13.15

Table 21.1   *(Continued)*

Age Interval (Years)[a]	Proportion Dying: Proportion of Persons Alive at Beginning of Age Interval Dying During Interval	Of 100,000 Born Alive		Average Remaining Lifetime: Average Number of Years of Life Remaining at Beginning of Age Interval
		Number Alive at Beginning of Age Interval	Number Dying During Age Interval	
68–69	.03301	65,267	2,155	12.55
69–70	.03573	63,112	2,255	11.96
70–71	.03866	60,857	2,352	11.38
71–72	.04182	58,505	2,447	10.82
72–73	.04530	56,058	2,539	10.27
73–74	.04915	53,519	2,631	9.74
74–75	.05342	50,888	2,718	9.21
75–76	.05799	48,170	2,794	8.71
76–77	.06296	45,376	2,857	8.21
77–78	.06867	42,519	2,920	7.73
78–79	.07535	39,599	2,983	7.26
79–80	.08302	36,616	3,040	6.81
80–81	.09208	33,576	3,092	6.39
81–82	.10219	30,484	3,115	5.98
82–83	.11244	27,369	3,078	5.61
83–84	.12195	24,291	2,962	5.25
84–85	.13067	21,329	2,787	4.91
85–86	.14380	18,542	2,666	4.58
86–87	.15816	15,876	2,511	4.26
87–88	.17355	13,365	2,320	3.97
88–89	.19032	11,045	2,102	3.70
89–90	.20835	8,943	1,863	3.45
90–91	.22709	7,080	1,608	3.22
91–92	.24598	5,472	1,346	3.02
92–93	.26477	4,126	1,092	2.85
93–94	.28284	3,034	858	2.69
94–95	.29952	2,176	652	2.55
95–96	.31416	1,524	479	2.43
96–97	.32915	1,045	344	2.32
97–98	.34450	701	241	2.21
98–99	.36018	460	166	2.10
99–100	.37616	294	111	2.01
100–101	.39242	183	72	1.91
101–102	.40891	111	45	1.83
102–103	.42562	66	28	1.75

Table 21.1   (Continued)

Age Interval (Years)[a]	Proportion Dying: Proportion of Persons Alive at Beginning of Age Interval Dying During Interval	Of 100,000 Born Alive		Average Remaining Lifetime: Average Number of Years of Life Remaining at Beginning of Age Interval
		Number Alive at Beginning of Age Interval	Number Dying During Age Interval	
103–104	.44250	38	17	1.67
104–105	.45951	21	10	1.60
105–106	.47662	11	5	1.53
106–107	.49378	6	3	1.46
107–108	.51095	3	2	1.40
108–109	.52810	1	0	1.35
109–110	.54519	1	1	1.29

[a]First entry in first row represents interval from birth to first birthday.
Source: U.S. National Center for Health Statistics, Life Tables, 1959–61, Vol. 1, No. 1, "United States Life Tables: 1959–61," December 1964, pp. 8–9.

before they reach their twenty-first birthday. Thus, an individual celebrating his twentieth birthday has roughly 1 chance in a 1,000 of dying before reaching his twenty-first birthday (the more precise chance is 1.15 in 1,000). Actuaries and public health specialists use the term **hazard rate**, or the more old-fashioned and sinister-sounding term **force of mortality** for this chance. The entry (96,111) in the third column represents the number of individuals who survived until their twentieth birthday out of an original group of 100,000 individuals alive at birth; this means, of course, that of this group of 100,000, 3,889 (100,000-96,111) individuals died before reaching their twentieth birthday.

The entry (110) in the fourth column represents the number of people who died between their twentieth and twenty-first birthdays; just as in column 3, we start with an original group of 100,000 people alive at birth. The individuals in this (fictitious) group of 100,000 born at the same time are called **cohorts**. The observant reader might wonder about the apparent discrepancy between the rate of death (115 per 100,000) in column 2 and the rate of death (110 per 100,000) in column 4. No discrepancy really exists—the **hazard rate** in column 2 represents the proportion dying in the 20-21 year age group of *those alive at their twentieth birthday*; the rate in column 4 represents the proportion dying in the 20-21 year age group of *the original* *group born alive.* The rates differ simply because the denominators differ. Finally, the entry (52.58) in the last column shows the remaining number of years of life an individual may expect to live on the average as he celebrates his twentieth birthday. Of course, a particular person just reaching age 20 may live only 1 more year or may live 65 more years—the value 52.58 represents the **average** number of years left to him. In other words The statistician sums up the numbers of remaining years of life of the 96,111 individuals just reaching age 20 and divides by 96,111.

We have illustrated the entries in the table for individuals in the 20–21 year age group. In a similar fashion, we may obtain comparable information for any yearly age group starting from birth and ending at age 110. Note that during the year 110, the last of the original group of 100,000 alive at birth dies.

We have simplified our description of the entries of a life table in that we have neglected to explain how one takes into account the fractional years of life lived by those who die in any year. The interested reader should

"Please check your entry list. The life table shows that I have 10 more years of mean remaining life."

consult Chapter 9 of Chiang (1968) for details as to how fractional years are accounted for in the calculations.

What uses can be made of a life table such as displayed in Table 21.1?

A. An insurance company clearly needs such information to arrive at the insurance premium charged corresponding to the age of a new purchaser of life insurance. Actually, insurance companies use life tables containing much more detailed information. For example, their rates will depend not only on the age of the individual, but also on certain other aspects: (1) male or female, (2) occupation, (3) health history, (4) marital status, and so on. These aspects influence the number of years of life remaining for the individual.

B. Pension planners use life tables to determine the monthly check appropriate to send to a retired person based on relevant factors, such as years of active contributions into the fund, salary earned, age at retirement, sex of retired person, and so forth. In this context, pensions refer to the Social Security Fund, industrial pensions, government workers' pensions (which include the enviable feature that the rising cost of living is taken into account), annuity funds, and so on.

In all of these cases, the **actuary** (a statistician specializing in the distribution of life length and its applications) uses life tables, but undoubtedly containing much more detail than the relatively simple one presented in Table 21.1.

C. Governmental planners, industrial market specialists, economists, and a variety of other analysts use life tables. They may be especially interested in how the life table changes as time passes. For example, planners of hospital facilities take into account the increasing proportion of older people in the population. This shift in population composition also influences a great variety of other planners: the makers of Geritol, the builders of retirement homes, the legislators setting the mandatory retirement age, the builders of residential dwellings of various types, and so on.

Summary

1. Life tables give the survival experience of a group of individuals (also fruit flies, patients, and so on).
2. The life table is typically divided into age intervals. Important quantities that can be obtained from the life table include, for each age interval:
   a. The hazard rate (the chance of dying in that age interval given that you have survived up to that interval)
   b. The proportion dying in that interval, among the original group born alive

   c. The average number of years left to a person who has just entered
      that age interval
3. Among the many enterprises and groups who find life tables extremely
   valuable are insurance companies, pension planners, the building
   industry, biologists studying specific species, and so forth.

Problems

1. For a 58-year-old person, what is the average number of years of life
   left? Since the life table in the text is over 20 years old, is the entry more
   likely to be too low or too high?
2. Your spouse dislikes you on your twenty-eighth birthday (at other times
   too). What is the chance that you will die during the next year? How
   many additional years of wedded misery can your spouse dread
   (assuming both parties stick it out and your spouse outlives you)?
3. A newborn babe has approximately 26 times as much chance of dying
   during the first year as compared with the chance that a fresh 2-year-old
   (they usually are) has of dying during the year ahead! Verify and then
   explain (to the mother, not to the children).
4. Notice that the average number of years of life remaining does not
   diminish by 1 as a person gets older by 1 year. Explain.
5. During which age interval does the largest number of people die? This
   interval is called the **modal** interval.
6. Note that the hazard rate during the age interval 0-1 is about the same as
   during the age interval 65-66. What factors explain the phenomenon?
7. During what age interval is the hazard rate lowest?
8. Suppose, on the average, an insurance company pays back as many
   dollars to the surviving family as it received in premiums over the many
   years during which the insurance was in force. What economic and
   financial factors make this a profitable arrangement for the insurance
   company? Hint: How much can the dollar buy at the time the premium is
   paid to the insurance company? How much can it buy when the
   insurance company pays back the surviving family? How does the
   insurance company use the money it receives over the many years before
   paying back the survivors? (Would you rather "own a piece of the Rock"
   or own the stock of the Rock?)
9. Why are the age intervals of Table 21.1 in years rather than months?
   Can you think of a situation in which months are preferable? (Please
   allow yourself the possibility of considering nonhuman populations.)

References

1. C. L. Chiang (1968). *Introduction to Stochastic Processes in Biostatistics*. Wiley, New York.
2. A. J. Gross and V. A. Clark (1975). *Survival Distributions: Reliability Applications in the Biomedical Sciences*. Wiley, New York.

Although references 1 and 2 are mathematical, they may be of interest to the reader who wants to learn more about life tables and other problems concerning the expected time to some end-point event.

Vignette 22

Estimating Survival Probabilities

Klaus painfully recalled the doctor's exact words: "Based on experience, I'd say that you have a 50–50 chance of living an average of at most six months, give or take a month. I wish I could be more precise, but your case is unique, to my knowledge."

Klaus furrowed his brows as he struggled with the doctor's grim words. His difficulty in fully understanding their meaning, in spite of the statistics course he had completed at the University of Berlin, he decided morosely, was just another telltale symptom of his progressively deteriorating diseased brain. He gave up the struggle and angrily snapped on the TV.

Manfred Herst, *Death Row*

Table 22.1, based on a study by Frank Proschan (1963), gives times to first failure of air-conditioning equipment in 13 Boeing 720 airplanes.

A new Boeing 720 is being put into service. Think of it as plane #14. How can we estimate the chance that its air-conditioning equipment will survive at least 400 operating hours before failing?

We may reason as follows. Table 22.1 shows that the air-conditioning equipment of two planes (#2 and #12) survived at least 400 hours, whereas the air-conditioning equipment for the other 11 planes survived less than 400 hours. Thus, it is reasonable to estimate the chance that plane #14 will accumulate at least 400 failure-free hours before the first failure of its air-conditioning equipment to be 2/13 = .15, namely, the observed proportion of survivals of at least 400 hours.

Table 22.1 Operating Hours Until First Failure of Air-Conditioning Equipment in 13 Airplanes

Airplane	1	2	3	4	5	6	7	8	9	10	11	12	13
Hours	194	413	90	74	55	23	97	50	359	50	130	487	102

Similarly, suppose we want to estimate the chance that the air-conditioning equipment of plane #14 will survive at least 100 hours. Our estimate is $6/13 = .46$ because among planes #1-13, six have survival times of at least 100 hours (planes #1, 2, 9, 11, 12, and 13).

A graph displaying the observed fraction surviving (empirical survival function) can be helpful in estimating survival probabilities. The horizontal axis of the graph represents time (operating hours), and the vertical axis represents observed fraction surviving. To draw this graph, we first put the observed survival times in order of increasing size. The ordered survival times are: 23, 50, 50, 55, 74, 90, 97, 102, 130, 194, 359, 413, and 487. Starting at time 0, we begin a horizontal line at a height of 1. This line continues until the smallest survival time of 23. At 23, we begin a new horizontal line at a height of $12/13 (=1 - 1/13)$. This line continues up to the next failure time 50. At 50 we begin a new horizontal line at a height of $10/13 (=1 - 3/13)$, and this line continues until the next failure time 55. At 55, we begin a new horizontal line at a height of $9/13 (=1 - 4/13)$ and continue it up to the next failure time of 74. We continue this process. Each time we come to a new ordered survival time, the height of the horizontal line drops by a multiple of $1/13$. The drop is $1 \times 1/13$ if there is only one failure at that time. The drop is $2 \times 1/13 = 2/13$ if there are two failures at that time (as is the case for the failure time 50). More generally, at a given failure time, the height of the new horizontal line is reduced from the height of its successor by the amount: Number of failures at that given failure time $\times 1/13$.

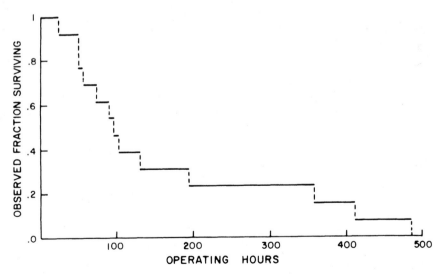

Figure 22.1   The empirical survival function for the air-conditioning system.

At the time of the last failure, the height of the horizontal line reaches 0 and remains at 0 thereafter. This sequence of horizontal lines, the heights of which decrease as one moves to the right along the time axis, depicts the empirical survival function. It is pictured in Fig. 22.1.

Suppose we wish to estimate the chance that the air-conditioning equipment of plane #14 will survive at least 200 hours before failing. Our estimate is read from the graph as the height 3/13 corresponding to the horizontal value 200.

By using airplanes #1–13 to estimate survival probabilities for airplane #14, we are assuming that the 14 planes represent a *homogeneous* group. We know that they are all of the 720 type, but other factors that influence the survival of the air-conditioning equipment may differ among the planes. These include operating conditions, time of day of flights, maintenance policies, and so on. Our estimates of survival probabilities for plane #14 tend to be better, the less the 14 planes differ with respect to variables that affect the survival of the air-conditioning system.

Summary

1. The empirical survival function (observed fraction surviving) is useful for estimating survival probabilities.
2. When graphing the empirical survival function, use the horizontal axis to represent time (operating hours, days, years, and so on) and the vertical axis to represent empirical survival probability.

Problems

1. Using the data of Table 22.1, estimate the chance that the air-conditioning equipment of plane #14 will fail during the first 100 operating hours.
2. Note that the largest observation in Table 22.1 is 487. At this observation, the empirical survival function has a height of zero. Thus, our estimate of the chance that "plane #14 will operate at least 487 hours before failing" is zero. Do you think this estimate is reasonable? Can you think of a way of improving it?
3. Note that the smallest observation in Table 22.1 is 23. Since none of planes #1–13 had air-conditioning equipment failure before 23 operating hours had elapsed, we estimate the chance that plane #14 will operate at least 23 hours before air-conditioning failure to be 1. Is this reasonable? How can you improve this estimate?
4. Suppose we tell you that the air-conditioning equipment of plane #14 has already operated 100 hours without failure. Using this information,

how would you estimate the chance that it will operate an additional 400 hours without failure?

5. The following data are from a study of the effect of methylmercury poisoning on survival time of goldfish. For a dosage of .001 ppm, the ordered survival times (in hours) are: 42, 43, 51, 61, 66, 69, 71, 81, 82, and 82 for 10 unhappy goldfish. If an eleventh goldfish is subjected to these conditions, what is your estimate that it will survive at least 65 hours?

6. Draw the empirical survival function for the data in Problem 5.

7.* Recall that you have already been introduced to the topic of estimating survival probabilities in Vignette 21, "Life Tables." In that vignette our estimates are based on large numbers of **grouped observations.** A table of grouped observations lists successive intervals and shows the number of observations that fall into each interval; the exact value of an observation is not shown. For example, from Table 21.1 of Vignette 21, we see that 171 people died between their thirty-fourth and thirty-fifth birthdays. Grouping is often used for convenience (it would be tedious to list 100,000 lifetimes in terms of days lived), or because more accurate observations are not available.

How would you plot an empirical survival function for the data of Table 21.1, Vignette 21?

8. How would you plot a graph which estimates the chance of *not* surviving up to a specified time for any specified time?

9. The **median** of a sample is the value that divides the sample into two halves. For a sample containing an odd number of observations, one way to find the median is to list the observations from least to greatest and determine the middle one. (For example, if the sample is 42, 12, 18, then the ordered sample is 12, 18, 42, and the middle one in the ordered list is 18; this is the sample median.) If the sample consists of an even number of observations, the median is taken to be the average of the two middle observations. (For example, in Problem 5 the sample median is $(66 + 69)/2 = 67.5$.)

Suppose we do not let you see the sample, but you can see the empirical survival function. How would you determine the sample median?

10. Refer to the grouped observations of Table 21.1, Vignette 21. How would you estimate the probability of dying by age 41½, given that you survived up to age 41?[†]

[†] M. H., you do make up some sick problems! If I were of age 41, I'd avoid this problem like the plague.   F. P.

Reference

1. F. Proschan (1963). Theoretical explanation of observed decreasing failure rate. *Technometrics* 5, 375–383.

Objectively viewed, this is a great paper. It is a fascinating blend of statistical theory and practice, applied to a real-life situation.   F. P.

Vignette 23
Reliability Growth

A new weapon system is being developed. At successive stages of development, tests are conducted to determine the proportions of failures of two basic kinds:

A. **Inherent** failures, that is, failures whose removal would require a significant improvement in the state of the art.

B. **Assignable-cause** failures, that is, failures whose cause can be identified; having identified the cause, the engineer can make a design change which results in improved reliability of the system.

The information accumulated under a testing program of this type has several basic uses:

1. Continual improvement of the system by weeding out successively the various types of assignable-cause failures discovered, as mentioned in B above.

2. Estimation of the probability that a trial will result in failure of the inherent type.

3. Estimation of the probability that a trial will result in failure of the assignable-cause type.

4. Estimation of the probability that a trial will result in successful functioning of the system; we call this probability the **reliability** of the system.

Several facts are immediately apparent.

a. The probability of an inherent failure does *not* change as the development program progresses, since no remedial action can be taken against this type of failure.

b. The probability of failure due to an assignable cause *decreases* as the development program progresses. Obviously, as assignable causes of failure are discovered and eliminated, the probability of failure resulting from assignable causes not yet discovered will decrease.

c. System reliability *increases* as the development program progresses, because of the improvements mentioned under fact b.

We illustrate (see Tables 23.1 and 23.2) with a set of real data for an important weapon system how the statistician estimates the probability of an inherent failure, the probability of an assignable-cause failure at successive stages of development, and the system reliability at successive stages of development. Of intense interest to the program manager is the *current* system reliability. (The identity of the system is not specified for obvious reasons.)

First, we note that out of the 54 trials that the system has undergone, a total of 10 inherent failures were observed. Thus, we estimate that the probability of an inherent failure is $10/54 = .1852$. Of course, this is an **estimate**, based on a modest number of trials. The unknown true probability of inherent failure will undoubtedly differ somewhat from this estimate. Various statistical methods are available for making precise **probabilistic** statements about the discrepancy between the estimate (.1852) and the unknown true value. Clearly, the size of the discrepancy will tend to decrease with the total number of trials conducted (**sample size**). A crude rule of thumb is that the discrepancy will be roughly half as large if we increase our sample size by a factor of 4, one-third as large if we increase our sample size

Table 23.1 Data Observed in Successive Stages

Stage	Number of Trials	Number of Inherent Failures	Number of Assignable-Cause Failures	Number of Successes
1	1	0	1	0
2	1	0	1	0
3	1	0	1	0
4	3	1	1	1
5	5	0	1	4
6	1	0	1	0
7	1	0	1	0
8	4	0	1	3
9	37	9	1	27
Total	54	10	9	35

by a factor of 9, and so on. Thus, the error of the estimate is roughly inversely proportional to the square of the sample size.

Next we wish to estimate the probability that a single trial at the first stage results in an assignable-cause failure *assuming that the outcome is not an inherent failure.* We shall be excluding inherent failures in estimating this conditional probability of assignable-cause failures, since we have already estimated the proportion of trials resulting in inherent failures (.1852).

At the first stage, one assignable-cause failure occurred in one trial; no inherent failure occurred. Thus, our "raw" estimate of the chance of assignable-cause failure among noninherent failure trials *at stage 1* is $1/1 = 1$. An important point should be brought out now. Since we are constantly improving the system by removal of assignable-causes of failure, we expect a priori that the chance of an assignable-cause failure will steadily diminish during successive stages of testing. However, due to sampling fluctuations, the *observed* (as distinguished from the unknown *true*) chance of an assignable-cause failure may not always diminish. We shall explain a statistical procedure for adjusting the observed data to take into account the engineering improvement occurring from stage to stage.

In Table 23.2 we display the raw data and the adjusted data relevant for the calculation of the chance of an assignable-cause failure when inherent failures are excluded from consideration.

The entries in the first 4 columns are clearly explained by the column headings. In column 5, we enter *on the left side* the ratio of the observed number of assignable-cause failures (from column 2) to the observed number of trials excluding inherent failures (from column 4). *On the right side* of column 5, we write the simplified form of the ratio (e.g., 1/1 in stage 1 reduces to 1); we also mark by a curly bracket the first occurrence of a reversal in the sequence of nonincreasing values. Note that the first five values, 1, 1, 1, 1/2, 1/5 *are* nonincreasing. However, between stage 5 and stage 6 a reversal occurs since the stage 5 value, 1/5 is strictly *smaller* than the stage 6 value, 1, thus constituting a reversal in the sequence of nonincreasing probabilities expected in this column. This reversal is the result of sampling fluctuations. More specifically, the unknown true proportion of trials excluding inherent failures which result in assignable-cause failure *is* nonincreasing. However, since we are observing only a small sample at each successive stage, it is not surprising that now and then the *observed proportion* will shown an increase (rather than the usual decrease) simply due to chance.

We may readily eliminate the reversal between stages 5 and 6 by **pooling** the data in the two stages. Specifically, we add the numbers of assignable-cause failures in stages 5 and 6 to obtain $1 + 1 = 2$ assignable-cause failures: similarly, we add the numbers of trials excluding inherent

Table 23.2 Estimation of Assignable-Cause Failure Probabilities

Column 1	2	3	4	5	6	7
Stage	Number of Assignable-Cause Failures	Number of Successes	Number of Trials Less Number of Inherent Failures	Initial Estimate of Chance of Assignable-Cause Failure Assuming Inherent Failure Does *Not* Occur	Adjustment to Eliminate First Reversal in Column 5 Probabilities	Adjustment to Eliminate First Reversal in Column 6
1	1	0	1	$1/1 = 1$	$1 = 1$	1
2	1	0	1	$1/1 = 1$	$1 = 1$	1
3	1	0	1	$1/1 = 1$	$1 = 1$	1
4	1	1	2	$1/2 = 1/2$	$1/2 = 1/2$	1/2
5	1	4	5	$1/5 = 1/5$ [a]	$2/6 = 1/3$	3/7
6	1	0	1	$1/1 = 1$	$2/6 = 1/3$ [b]	3/7
7	1	0	1	$1/1 = 1$	$1 = 1$	3/7
8	1	3	4	$1/4 = 1/4$	$1/4 = 1/4$	1/4
9	1	27	28	$1/28 = 1/28$	$1/28 = 1/28$	1/28
Total	9	35	44			

[a] First reversal in column 5.
[b] First reversal in column 6.

failures to obtain $5 + 1 = 6$. We then form a new ratio $2/6 = 1/3$ as our revised estimate of the chance of an assignable-cause failure assuming inherent failures do not occur. This revised estimate $1/3$ now applies to *each* of stages 5 and 6, and is so recorded in column 6. Thus, in column 6 we now have a nonincreasing sequence of estimates at least down to stage 6.

Continuing on in column 6, we encounter another reversal in going from stage 6 to stage 7: $1/3$ is *smaller* than 1. To eliminate this reversal, we must exercise care since there are *two* values of $1/3$ (stages 5 and 6) and one value of 1 (stage 7). To eliminate the reversal involving stages 5, 6, and 7, we pool the relevant data from the three stages. We thus obtain as our revised estimate of the chance of an assignable-cause failure assuming inherent failures do not occur, the ratio, $3/7$, which applies to each of stages 5, 6, and 7. (In stages 5, 6, and 7, there is a total of 3 assignable-cause failures, while the total number of trials excluding inherent failures is 7.) Entering the common value $3/7$ in stages 5, 6, and 7 of column 7, we note that the resulting sequence is now completely free of reversals; that is, the successive entries in column 7 are nonincreasing as we read down the column.

These entries now constitute our estimates of the chances of assignable-cause failure assuming no inherent failure occurs for the successive stages of system development.

We may now estimate for the system the chance of an inherent type of failure, the chance of an assignable-cause type of failure, and the reliability; moreover, these probability estimates will reflect the improvement in system reliability resulting from elimination of assignable-causes of failure after successive stages of development.

We have already estimated the chance of an inherent type of failure as .1852. By definition, this chance does not change as the system improves. In column 7, we have obtained an estimate of the chance of an assignable-cause failure assuming inherent failure does not occur, taking into account improvements with system development. Thus, we can estimate reasonably well the *current* system reliability as

$$1 - .1852 - (1/28)(1 - .1852) = .7857.$$

This estimate is obtained by subtracting from 1 (perfect reliability) (1) the proportion .1852 of inherent failures observed in the 54 trials and (2) the updated chance of an assignable-cause type of failure. The second subtraction requires some explanation. The factor $1/28$ is the estimate of the current chance of an assignable type of failure assuming no inherent failure; the factor $(1 - .1852)$ multiplying $1/28$ represents the chance of no inherent failure. The product therefore represents the current chance of an assignable-cause failure. The final answer .7857 is our estimate of current system reliability (that is, after nine stages of testing and successively removing various types of assignable-cause failures).

There are several advantages to this method of estimation. First, we realistically distinguish between the two basic types of failure: inherent failure and assignable-cause failure. Second, we obtain an estimate of the chance of each type, using our prior knowledge that the chance of an assignable-cause failure will decrease as successive stages of the testing program occur. Finally, we estimate the *current* reliability of the system, again incorporating our prior knowledge that the reliability of the system is increasing as assignable-cause failures are detected in successive stages of testing and removed.

The reader may ask: What would be our estimate of reliability if we did not distinguish between inherent and assignable-cause failures and did not make use of the prior information that system reliability is improving through removal of the causes of assignable-cause failures? Such an estimate would be $35/54 = .6481$, the 35 representing the total number of successful trials in the 9 stages, the 64 representing the total number of trials. Note that this (incorrect) estimate is distinctly lower than the more realistic estimate .7857 which takes into account the known physical fact that reliability is improving from stage to stage. This difference in system reliability estimates would undoubtedly have very important consequences to both the contractor and the government. It could result in savings of millions of dollars and perhaps months of time that might have been expended in unrealistic efforts to improve system reliability, especially at later stages of system development.

Summary

1. In the stages of development of a system, it is useful to distinguish between inherent failures (failures where removal would require a significant improvement in the state of the art) and assignable-cause failures (failures whose cause can be identified and, after identification, design changes can be made which improve reliability of the system).
2. The probability of an inherent failure does not change as the development program progresses because no remedial action can be taken for this type of failure.
3. The probability of failure due to an assignable cause decreases as the development program progresses. As assignable causes of failure are discovered and eliminated, there is less chance of failure due to assignable causes not yet discovered.
4. The reliability of the system increases as the development program progresses, due to the improvements mentioned in 3. Such an increase in reliability is termed reliability growth.

Problems

1. If the inherent type of failure did *not* exist (that is, the cause of every failure could be identified and removed), would the system reliability estimate at the last (ninth) stage be higher? Explain.

2. Note that there is exactly one assignable-cause failure per stage. Why?

3. If the number of trials per stage were considerably larger, would you expect fewer reversals for a fixed number (say nine, as in the present example) of stages? Explain.

4. A high school physical education teacher measures the heights of the boys in his freshman class every 2 months. The average measured heights in succession are:

Measurement number	1	2	3	4	5	6
Average measured height, in.	63.4	63.6	63.5	63.7	63.8	64.0

   a. Note the reversal between measurement average 2 and measurement average 3. What values would seem more reasonable in view of the fact that boys tend to grow taller as they get older?

   b. If the teacher had recorded the average height based on his four freshman classes, would such a reversal be less likely? Explain.

5. In the reliability growth example discussed in the text, we noticed that in column 6 of Table 23.2, although the reversal occurred only between stages 6 and 7, we pooled stages 5, 6, and 7. Why?

6. Explain why it is advantageous to the contractor designing the weapon system to use the statistical method described for estimating current reliability.

7.* A contractor is designing a new tank. After each test run of the experimental tank, he introduces design changes which improve the performance and lengthen the time until failure of the tank. The data for intervals between successive failures are shown in Table 23.3.

Table 23.3   Intervals Between Successive Failures of Experimental Tank

Interval number	1	2	3	4	5	6	7	8	9	10
Length of interval, hours	29	40	38	54	50	60	72	70	71	80

   a. Estimate the lengths of intervals between failures assuming tank improvement after each test run.

   b. What is your estimate of current (latest) interval between failures?
   c. Suppose an estimate of **mean time between failures** had been
      obtained as a simple **arithmetic average** of the 10 intervals. How
      would it compare with the estimate of current interval between
      failures assuming reliability growth, obtained in b? Why is it
      advantageous to the contractor to estimate the current interval
      between failures assuming reliability growth?
   d. Suppose design changes cease after the last (tenth) test run shown.
      How can the contractor obtain a more reliable estimate of current
      time between failures?
 8.* Refer to Problem 7. Suppose in Table 23.3 the observation for interval
      #5 were missing. What would be a reasonable way to solve part d in
      the absence of the observation?
 9. Give an example of reliability growth in a human system.
 10. Can you devise a system development program where reliability growth
      is not possible? Explain.

Reference

1. R. E. Barlow and E. M. Scheuer (1966). Reliability growth during a development testing program. *Technometrics 8*, 53–60.

Reference 1 has a nice easily understandable title. After that, it becomes technical,
mathematically. It is a useful and influential paper. If you happen to be a large
multibillion dollar contractor dealing with the government, you will want your project
directors to have their statisticians explain it to them. The project directors will then
be in a better position to claim high reliability for the systems they are developing.

Vignette 24

Revising Opinions

Shots rang out beyond the tent, barking and then whining, spanking the sand in spurts. . . . He calculated the odds that his identity had been detected, decided they were 50-50. . . .

John Updike, *The Coup*

"My theory of neurotic anxiety is surprisingly simple: The neurotic has a distorted set of probabilities concerning the risks he is exposed to. For example, the chance of a heart attack may be $1/1000$ for a person of his age; the neurotic may believe (feel?) that *his* chance is 1 in 2. The chance of an airplane crashing may be $1/100,000$; the neurotic aboard the plane *knows* in his pounding heart that the chance is really 1 in 3."

Professor Young paused for the admiring approval that he felt he had earned. (Clearly, *he* was not an anxious self-questioning neurotic.)

Ellen smiled, just as Professor Young had guessed she would. Her next remark thus came as a double shock.

"Professor Young, my evaluation of the chance that a statistician understands people has always been 1 in 3. After your brilliant reduction of the troubled human mind to a set of cold numbers, I now believe that the chance that a statistician understands or sympathizes with people is only 1 in 3,000,000. As your species might put it: Statisticians and people are mutually exclusive categories."

Ellen burst into tears and slammed out of the room.

Professor Young muttered to himself, "She has a great deal to learn about statistics and statisticians in general. Perhaps this particular statistician should teach her."

Ingram J. Carswell, *The Lady and the Statistician*

Sometimes we modify our opinion concerning the chance that a specific event will occur when we obtain additional information about the event. Consider some examples.

1. *The chance of rain:* You go to bed directly after accepting the 11:15 P.M. weather forecast that there is a 60% chance of rain tomorrow. When you wake up at 7:00 A.M. the sky is dense with ominous dark clouds and you hear thunder in the distance. Now you probably want to modify (undoubtedly, increase) your prior estimate of a 60% chance of rain.

2. *The chance of a hit:* Lance Starr plays major league baseball for the Los Angeles Dodgers. This year, in 390 official "at-bats," he has hammered 130 hits. In baseball jargon, Starr is called a .333 (130/390) hitter, or equivalently, he is said to have a .333 batting average. If we ask for your estimate of the chance that Starr will get a hit in his *next* official at-bat, it seems reasonable for you to reply 1/3. However, suppose we now give you the following additional information. In the next game (during which Starr is likely to record his next official at-bat), Los Angeles plays the San Francisco Giants. Sal Magnet, the Dodger "killer" and Lance Starr's personal nemesis, has been named as the Giant pitcher. This year Starr has had only 1 hit in 10 official at-bats against Magnet. It seems reasonable, in light of this additional information, to revise (specifically lower) your estimate of 1/3.

3. *The chance of disease:* Dr. Clark has a medical practice in a section of New York City where the **prevalence rate** of tuberculosis (TB) is approximately .001. Thus, Dr. Clark's estimate is .001 that a person who comes to him for a routine checkup will actually have TB. In such checkups, it is Dr. Clark's custom to order a chest X-ray for use as a diagnostic aid. Suppose the X-ray is positive, *indicating* that the person has TB. As we shall see in a moment, such a positive result does not make it certain that the person has TB. However, the positive test result will undoubtedly lead Dr. Clark to revise upwards from .001 his estimate of the probability that the person has TB.

It is clear from these examples that additional relevant information should lead to reconsideration and revision of initial probabilities. How does one use the observed data to convert the initial probability (called the **prior probability**) to the revised probability (called the **posterior probability**) computed *after* the additional data is observed?

Using additional information to revise or update prior probabilities falls in the realm of Bayesian statistical methods. Such methods are so named in honor of a mid-eighteenth century philosopher, Rev. Thomas Bayes.

Bayesian methods are clear in principle, but they may be difficult to use in practice. A careful exposition of the Bayesian approach to statistics is beyond the level of this book. Instead, we will use the diagnostic testing situation to present a simplified illustration of how probabilities can be reasonably revised.

If the diagnostic test were infallible, then revision of probabilities would be easy. *In such an ideal world,* if we observe a positive result on the test, we would revise the probability of TB from the prevalence rate to 1. Similarly, if we observe a negative result on the test, we would revise the probability of TB from the prevalence rate to 0.

However, the diagnostic test is not infallible. In fact there are two kinds of errors that the diagnostic test based on the chest X-ray makes. The test might be negative even though the person really has tuberculosis; thus, it erroneously indicates that the person is healthy. Such an incorrect result is called a **false-negative** result. The test might be positive even though the person does not have TB; thus, it erroneously indicates the healthy person has TB. Such an incorrect result is called a **false-positive** result.

The chances of false-negative or false-positive results depend to some degree on the skill of the X-ray reader. A less experienced reader may incorrectly interpret certain artifacts or transient lung abnormalities appearing in the X-ray as evidence of TB. On the other hand, the reader may miss actual evidence of TB.

There are ways of estimating a particular X-ray reader's error rates. For example, a reader may be presented in random order X-rays of patients *known* to have TB and X-rays of patients *known* to be free of TB. Then the reader is asked to classify the X-rays as to whether or not they indicate TB. From these results the reader's error rates can be directly measured.

Yerushalmy, Harkness, Cope, and Kennedy (1950) studied errors made by six readers in interpreting X-rays for TB. Table 24.1 contains the results for one of the six readers.

In 1,790 X-rays of persons without disease, the X-ray reader committed 51 false-positive errors, yielding a false-positive error rate of $51/1790 = .028$, or about 3%. For the 30 X-rays of people having TB, there were 8 false-negative errors, a false-negative error rate of $8/30 = .267$, or about 27%.

Suppose we selected an X-ray at random from the 1,820 X-rays categorized in Table 24.1. What is your estimate of the chance that the X-ray selected is that of a person with TB? You could reason as follows. Since 30 of the 1,820 X-rays correspond to actual cases of TB, the chance could be estimated as $30/1820 = .016$, or about 2%. (Note that this value of .016 is a

Table 24.1 Results in a Study of X-Ray Readings

	Persons Without TB	Persons with TB	Total
Negative X-ray reading	1,739	8	1,747
Positive X-ray reading	51	22	73
Total	1,790	30	1,820

bit too high to be a realistic value for the prevalence rate of TB for a typical group. But the data of Table 24.1 do not correspond to a typical group. The study of Yerushalmy *et al.* was performed to estimate error rates. In particular, to obtain a better estimate of the false-negative error rate, it was useful to have a proportion of diseased cases higher than found in a natural population.)

Now, suppose we tell you that the X-ray selected at random from the 1,820 available was scored positive by the X-ray reader. How should you revise your estimate of the probability of it being an X-ray of a person who actually has TB? Since out of 73 positive readings, only 22 persons had TB, a good choice for this revised probability would be 22/73 = .301, or about 30%. Thus, the additional knowledge that the X-ray was scored positive leads you to *revise* your estimate of the probability that the X-ray is that of a person who actually has TB from .016 to .301.

Summary

1. Additional information may lead to a modification of one's opinion about the chance that a specific event will occur.
2. The use of data to convert prior probabilities to posterior probabilities falls in the realm of Bayesian statistical methods.

Problems

1. An X-ray is selected at random from the 1,820 represented by Table 24.1. You are then told that this X-ray reading is negative. What is your revised estimate of the probability that the X-ray selected is that of a person with TB?

2. Savage (1968), using data reported by Rubin (1965), considers probabilities regarding missing persons. Table 24.2 categorizes by age and sex missing persons reported to the Cincinnati Police Department (1963) and to the Los Angeles Police Department (1964).

   Suppose a missing person is selected at random from the 8,153 missing people categorized in Table 24.2.

   a. What is the prior probability that the person selected is a female?
   b. Suppose you have the further information that the randomly selected person was reported to the Cincinnati Police Department as a missing person. What is the posterior probability that the person is a female?
   c. Suppose you have the information that the person selected was between 35 and 44 in age and was reported to the Cincinnati Police Department. Now what is your revised estimate of the probability that the person selected is a female?

Table 24.2 Missing Persons Reported to Cincinnati and Los Angeles Police Departments

	Cincinnati (1963)			Los Angeles (1964)		
Age	Total	Male	Female	Total	Male	Female
Under 18	1,246	644	602	5,250	2,447	2,803
18–24	178	92	86	194	58	136
25–34	218	143	75	172	96	76
35–44	157	113	44	161	88	73
45 and over	217	168	49	360	218	142
Total	2,016	1,160	856	6,137	2,907	3,230

3. How would you assess the probability that Lance Starr will get a hit in his next official at-bat against Sal Magnet?

4. One famous statistician has *jokingly* remarked that when a weatherperson states there is a 60% chance of rain tomorrow, the weatherperson means that six people think it will rain while four people think that it will not rain. Is this a reasonable way to assess the probability?

"The chance of rain tonight is 60%."

5. Refer to Table 16.1 of Vignette 16, "Was the 1970 Draft Lottery Fair?" M. Hollander's birthday is in the month of March.
   a. What is the chance that his birthday received one of the numbers 1, 2, . . . , 100 in the draft order?
   b. Suppose you have the further information that M. Hollander's birthday occurs within the first 22 days of March. What is your revised estimate of the probability that his birthday received one of the first 100 numbers in the draft order?

6. If you are told that the entry in the northwest corner of a page of random numbers (see Vignette 12) is a 1, should you then revise your opinion about the chance that the entry in the southeast corner of that page is a 1? Explain.

7. Give two examples from your own experience of formally or informally revising your assessment of the probability of an event in the light of additional information.

8. Without looking ahead (can you restrain yourself?), what is your estimate of the probability that the next vignette exceeds 10 pages in length? Explain the basis for your calculation. Now look ahead. What is your revised estimate of the probability?

9. Suppose a woman is selected at random from the 14 women represented by Table 18.1 of Vignette 18. What is the chance that the woman's second choice for leisure-time companions is "both sexes?" Given that the woman selected lists her most preferred choice as "women," what is the chance that her second choice is "both sexes?"

References

1. E. Rubin (1965). On missing persons and missing tabulations. *The American Statistician 19*, 33–36.[†]

2. I. R. Savage (1968). *Statistics: Uncertainty and Behavior.* Houghton Mifflin, Boston.

3. J. Yerushalmy, J. T. Harkness, J. H. Cope, and B. R. Kennedy (1950). The role of dual reading in mass radiography. *American Review of Tuberculosis 61*, 443–464.

References 1 and 3 contain the data of Tables 24.2 and 24.1, respectively. Reference 2 is of interest because it uses elementary gambling considerations to derive basic rules for computing and revising probabilities.

[†] This issue may be missing from your library.

Vignette 25
The Birthday Problem

The following problem has captivated the interests of many students in elementary probability and statistics classes. In a class of 28 students, what is the chance that at least 2 students have the same birthday (ignoring year of birth)?

Incorrect reasoning goes something like this. Excluding February 29, there are 365 possible birthdays but only 28 students. Thus, the chance should be about $28/365 = .077$, or about 8%. Actually the chance is much higher—about 65%. Many people find this result striking and hard to believe.

Before we justify it by precise calculation, we present similar probabilities for various-size classes. Table 25.1 gives the probability that at least two students have the same birthday and the complementary probability that no two students in the class have the same birthday. These probability values may seem counterintuitive, but we will prove that they are correct.

Note that when the number of students is 23, the chance is already slightly greater than 50% that at least 2 of the students in the class have the same birthday. That is, if you checked with a large number of groups consisting of 23 people each, about half of the groups would have birthday coincidences, while the other half would not.

When there are 64 in the class, it is *virtually certain* (the chance is .997) that the class will have at least two people with the same birthday.

We now describe how the values in columns 2 and 3 of Table 25.1 are obtained. First, for simplicity, we exclude February 29, so that there are only 365 possible birthdays. Second, we assume all 365 birthdays are equally likely. That is, if a student is selected at random from the class, we assume that each day of the year is equally likely to be that person's birthday. Now, we start with the simplest case: suppose that the class contains only *two* students. Since we have assumed all birthdays are equally likely, the probability that the two students have different birthdays is equal to:

Table 25.1   Birthday Coincidence Probabilities

Number of Students in Class	Probability that No Two Students in the Class Have the Same Birthday	Probability that Two or More Students in the Class Have the Same Birthday
2	.997	.003
4	.984	.016
8	.926	.074
12	.833	.167
16	.716	.284
20	.589	.411
22	.524	.476
23	.493	.507
24	.462	.538
28	.346	.654
32	.247	.753
40	.109	.891
48	.039	.961
56	.012	.988
64	.003	.997

$$\frac{\text{number of ways to assign birthdays to two students so that their birthdays are different}}{\text{total number of ways to assign birthdays to two students}}.$$

To evaluate the denominator of the fraction, note that we can assign the first birthday in 365 ways and the second birthday in 365 ways; thus, the total number of ways to assign two birthdays is 365×365.

To compute the numerator of the fraction, we count the number of ways to assign two birthdays so that the two birthdays are *not* the same. We still choose the first birthday in 365 ways, but since the second birthday must not be the same as the first birthday, there are only 364 ways to choose the second birthday. This gives a total of 365×364 ways to assign 2 birthdays so that they are not the same.

Thus, the desired probability is:

Probability that the 2 students in the class have different birthdays

$$= \frac{365 \times 364}{365 \times 365} = \frac{364}{365} = .997.$$

By similar reasoning, if the class contains exactly 3 students, then

Probability that the 3 students
in the class have different birthdays

$$= \frac{365 \times 364 \times 363}{365 \times 365 \times 365} = \frac{364 \times 363}{365 \times 365} = .992.$$

In a class with just 4 students,

Probability that the 4 students
in the class have different birthdays

$$= \frac{365 \times 364 \times 363 \times 362}{365 \times 365 \times 365 \times 365} = .984.$$

Continuing in this fashion, you could, with the aid of a good calculator, verify the entries in column 2 of Table 25.1. Futhermore you can calculate, for any specific number (not listed in the table) of students in the class, the chance that all the students have different birthdays.

To get a value in column 3 of Table 25.1, simply subtract the corresponding entry in column 2 from the number 1. For example, in row 1 find $.003 = 1 - .997$; in row 2, $.016 = 1 - .984$, and so on.

Recall that we started our discussion with exactly 28 students in the class. Thus,

Probability that the 28 students
all have different birthdays

$$= \underbrace{\frac{365 \times 364 \times 363 \times \ldots \times 338}{365 \times 365 \times 365 \times \ldots \times 365}}_{\text{28 factors of 365}} = .346,$$

and

Probability that two or more of the 28
in the class have the same birthday
$$= 1 - .346 = .654.$$

There are several reasons why students tend to underestimate this probability when they guess it.

First, a student may ask himself: What is the chance that out of the remaining 27 students at least 1 will have the same birthday as mine? He should really be asking: What is the chance that out of the 28 students, at least 2 will have birthdays that agree? (But they do not necessarily have to agree with his own birthday.)

"That's strange. In a class of five students, there is only a 3% chance of a birthday coincidence."

Second, a student, when thinking about how at least two people can have the same birthday, may neglect many possibilities. The possibilities include not only a pair of students having the same birthday, but also, to name another possibility, two students having one date in common *and* three others having another date in common, and so forth.

Summary

1. We have shown how to calculate the probability that in a class consisting of (for example) 28 students at least 2 have the same birthday.

2. The calculation of this probability utilizes an "equally likely" assumption that could be questioned.
3. To the inexperienced reader, the results of this vignette may seem counterintuitive. Intuition is often helpful, but in this case careful and rational thought, backed up by a correct probability calculation, will show that intuition has led one astray.

Problems

1. Table 25.2 gives the birthdays of the 39 presidents of the United States as of 1983. Note that James Polk and Warren Harding were both born on November 2, and that Andrew Johnson and Woodrow Wilson were both born on December 29. Do you find these coincidences surprising in view of the calculations of this chapter?
2. Find out the birthdates of the current 100 U.S. senators. Do you expect to find any birthday coincidences?
3. A small class contains six students. What is the chance that at least two have the same *birthmonth*?
4. Do you think the assumption that all 365 birthdates are equally likely is reasonable? Explain.

Table 25.2   Birthdates of American Presidents

George Washington	Feb. 22, 1732	Chester A. Arthur	Oct. 5, 1829
John Adams	Oct. 30, 1735	Grover Cleveland	Mar. 18, 1837
Thomas Jefferson	Apr. 13, 1743	Benjamin Harrison	Aug. 20, 1833
James Madison	Mar. 16, 1751	William McKinley	Jan. 29, 1843
James Monroe	Apr. 28, 1758	Theodore Roosevelt	Oct. 27, 1858
John Quincy Adams	July 11, 1767	William H. Taft	Sept. 15, 1857
Andrew Jackson	Mar. 15, 1767	Woodrow Wilson	Dec. 29, 1856
Martin Van Buren	Dec. 5, 1782	Warren G. Harding	Nov. 2, 1865
William H. Harrison	Feb. 9, 1773	Calvin Coolidge	July 4, 1872
John Tyler	Mar. 29, 1790	Herbert C. Hoover	Aug. 10, 1874
James K. Polk	Nov. 2, 1795	Franklin D. Roosevelt	Jan. 30, 1882
Zachary Taylor	Nov. 24, 1784	Harry S. Truman	May 8, 1884
Millard Fillmore	Jan. 7, 1800	Dwight D. Eisenhower	Oct. 14, 1890
Franklin Pierce	Nov. 23, 1804	John F. Kennedy	May 29, 1917
James Buchanan	Apr. 23, 1791	Lyndon B. Johnson	Aug. 27, 1908
Abraham Lincoln	Feb. 12, 1809	Richard M. Nixon	Jan. 9, 1913
Andrew Johnson	Dec. 29, 1808	Gerald R. Ford	July 14, 1913
Ulysses S. Grant	Apr. 27, 1822	Jimmy Carter	Oct. 1, 1924
Rutherford B. Hayes	Oct. 4, 1822	Ronald Reagan	Feb. 6, 1911
James A. Garfield	Nov. 19, 1831		

   5. Suppose there are 28 students in a class. What is the chance that at
       least two of the students in the class will have the same *deathday*?
   6. Refer to Problem 5. Do you think the assumption that all 365
       deathdates are equally likely is reasonable? Explain.
   7. Which of the following assumptions do you have more faith in?
       a. All 365 birthdates are equally likely.
       b. All 365 deathdates are equally likely.
              Explain your answer. (Hint: Do you have more control over your
       birthday or your deathday?)
   8.* What is the probability that the person who is president of the United
       States in 1996 will have a birthdate that is different from those listed in
       Table 25.2?
   9.* At the time of this writing, 35 of the 39 presidents listed in Table 25.2
       have died. Find out their deathdates. Do you expect any common
       deathdates among the 35?
  10.* Refer to Problem 9. Do you see any relationship between a president's
       birthdate and deathdate? Would you expect some relationship?
       Explain.
  11. In July 1983 the roster of the Atlanta Braves major league baseball
       team contained 28 players.
       a. *Without* looking at Table 25.3, calculate the chance of at least one
          birthday coincidence among the 28 players.
       b. Now look at Table 25.3 and revise your probability of at least one
          birthday coincidence.

Table 25.3 Birthdates of the 1983 Atlanta Braves

Name	Birthdate	Name	Birthdate
Steve Bedrosian	Dec. 6, 1957	Randy Johnson	June 10, 1956
Bruce Benedict	Aug. 18, 1955	Mike Jorgensen	Aug. 16, 1948
Tommy Boggs	Oct. 25, 1955	Rufino Linares	Feb. 28, 1955
Tony Brizzolara	Jan. 14, 1957	Craig McMurtry	Nov. 5, 1959
Brett Butler	June 15, 1957	Donnie Moore	Feb. 13, 1954
Rick Camp	June 10, 1953	Dale Murphy	Mar. 12, 1956
Chris Chambliss	Dec. 26, 1948	Phil Niekro	Apr. 1, 1939
Ken Dayley	Feb. 25, 1959	Larry Owen	May 31, 1955
Pete Falcone	Oct. 1, 1953	Pascual Perez	May 17, 1957
Terry Forster	Jan. 14, 1952	Biff Pocoroba	July 25, 1953
Gene Garber	Nov. 13, 1947	Rafael Ramirez	Feb. 18, 1959
Terry Harper	Aug. 19, 1955	Jerry Royster	Oct. 18, 1952
Bob Horner	Aug. 6, 1957	Claudell Washington	Aug. 31, 1954
Glenn Hubbard	Sept. 25, 1957	Bob Watson	Apr. 10, 1946

12. Table 25.4 contains the rosters of the Los Angeles Dodgers and San Francisco Giants baseball teams, as they were on April 25, 1982. There are twenty-five players on each roster. With each player, we have included his birthdate. Without peeking at the table, answer parts a, b, and c.

    a. What is the chance that at least two of the Dodgers will have the same birthdate? (that is, what is the chance of a birthday coincidence on the Dodgers?).

Table 25.4 Rosters of Los Angeles Dodgers and San Francisco Giants on April 25, 1982

Name	Birthdate	Name	Birthdate
		Dodgers	
Terry Forster	1/14/52	Ron Cey	2/15/48
Dave Goltz	6/23/49	Steve Garvey	12/22/48
Burt Hooton	2/07/50	Bill Russell	10/21/48
Steve Howe	3/10/58	Steve Sax	1/29/60
Alejandro Pena	6/25/59	Dusty Baker	6/15/49
Ted Power	1/31/55	Pedro Guerrero	6/29/56
Jerry Reuss	6/19/49	Jay Johnstone	11/20/46
Dave Stewart	2/19/57	Ken Landreaux	12/22/54
Fernando Valenzuela	11/01/60	Rick Monday	11/20/45
Bob Welch	11/03/56	Jorge Orta	11/26/50
Mike Scioscia	11/27/58	Ron Roenicke	8/19/56
Steve Yeager	11/24/48	Derrel Thomas	1/14/51
Mark Belanger	6/08/44		
		Giants	
Jim Barr	2/10/48	Duane Kuiper	6/19/50
Fred Breining	1/15/55	Johnnie LeMaster	6/19/54
Alan Fowlkes	8/08/58	Joe Morgan	9/19/43
Rich Gale	1/19/54	Joe Pettini	1/26/55
Al Holland	8/16/52	Reggie Smith	4/02/45
Gary Lavelle	1/03/49	Guy Sularz	11/07/55
Renie Martin	8/30/55	Jack Clark	11/10/55
Greg Minton	7/29/51	Chili Davis	1/17/60
Dan Schatzeder	12/01/54	Jeff Leonard	9/06/57
Milt May	8/01/50	Champ Summers	6/15/48
Jeff Ransom	11/11/60	Max Venable	6/06/57
Dave Bergman	6/06/53	Jim Wohlford	2/28/51
Darrell Evans	5/26/47		

b. What is the chance that at least two of the Giants will have the same birthdate?

c. What is the chance that "there will be a birthday coincidence on the Dodgers and a birthday coincidence on the Giants."?

Now look at Table 25.4.

d, e, f. Calculate the chances sought in parts a, b, and c, conditional on having looked at Table 25.4.

Vignette 26
The Law of Averages

Messenger said, "Can you work out any equations of probability of one hitting here?"

"No, sir. A hurricane has no memory. Like a coin. If a coin comes up heads fifty times, the odds on the next flip are still fifty-fifty, head or tail. But if you flip it ten thousand times, you'll get five thousand heads, plus or minus . . . ."

John D. MacDonald, Condominium

Shortly after they were married, one of Corde's academic friends had congratulated him, saying, "Do you remember that old piece of business from probability theory, that if a million monkeys jumped up and down on the keys of typewriters for a million years one of them would compose *Paradise Lost*? Well, you were like that with the ladies. You jumped up and down and you came up with a masterpiece."

Saul Bellow, The Dean's December

Let us ask you a simple question. Suppose you have a coin that appears fair; that is, after careful physical examination, you conclude that the coin is just as likely to come up heads as tails in a single toss. Your statistics professor who apparently believes you have as little to do as he, asks you to toss the coin 100 times. After the ninety-ninth toss, you have observed 56 heads and 43 tails. Do you believe that on the final toss:

1. A tail is more likely to appear than a head?
2. A head is more likely to appear than a tail?
3. A head is as likely to appear as is a tail, so that either outcome has a chance of 1/2?

A frighteningly large number of people answer yes to question 1! They invoke the so-called law of averages, which goes somewhat as follows: on the average, there should be about as many tails as heads if the coin is **fair**. In the 99 outcomes thus far, more heads than tails showed up. Thus, to achieve the equal balance between heads and tails called for by the law of averages, a tail is due; that is, the proportion of tails would move closer to the theoretical 1/2 called for by the law of averages. On the other hand, if a head showed up on the one-hundredth toss, the proportion of tails would move even further away from the 1/2 theoretically called for by the law of averages.

It may come as a surprise to many people that:

A. The law of averages as described just above is not correct at all.
B. If the coin is truly fair, the last of the 100 flips is just as likely to yield a head as a tail.
C. If the coin does have some minute physical imperfection, then indications are that a tail is *less* likely to appear on the last toss than is a head.

Let us try to justify these assertions, which to some readers will seem like heresy.

A, B. The law of averages as just described should be modified to constitute a *prediction* of what is likely to happen in a sequence of *future* tosses. The fact that 56 heads and 43 tails appeared in the first 99 tosses can have no affect on the outcome of the one-hundredth toss yet to be made. After all, the coin has no memory; it doesn't know what happened in the past 99 tosses. If indeed it is a fair coin, then on each outcome the chance of a head is the same as the chance of a tail, namely 1/2—this fact applies equally well to the one-hundredth toss about to be performed, regardless of whatever may have happened in the past.

There are statistical laws that predict (with statistical fluctuation allowed for) the future behavior in a sequence of tosses to be made. One law that represents a somewhat more accurate description than the law of averages stated above may be put as follows: For a fair coin, the chance is high that the proportion of heads is close to .5 in a large sequence of tosses. The greater the number of tosses to be made, the smaller is the chance that the proportion of heads will differ from the theoretical value .5 by at most any specified small amount.

Note that the statement is more subtle and complex than the original version of the law of averages originally stated.

It refers to a sequence of future tosses. It does not use the past to forecast the future. It is based on the assumption of a fair coin (as was the law of averages originally put forward). It also takes into account the number of trials to be performed. Moreover, as is usual for us weasel-worded statisticians, the statement allows for a margin of "error"—the margin tending to become smaller as the number of trials becomes larger. Finally, it

VIRGINIA SMOLIAR

"'Lore of Ravages'! Young man, I suggest a porno shop, not a respectable library."

is a statement concerning the chance of outcomes; it does not state definitively that any specified outcome or set of outcomes will definitely occur. (You may get the impression from all this that the statistician is afraid to come right out and make unequivocal statements. It would be more precise to say that in discussing events that depend on chance, the statistician can only be accurate in making statements that describe the chance of various outcomes. The statistician is not a soothsayer!)

C. Suppose the coin does have some minute physical irregularities. After all, a *perfectly* balanced coin is an ideal concept, more likely to be conceptualized in a statistics text than encountered in our real world of imperfections. Thus, the coin may actually have a probability of landing heads which is not precisely 1/2, but just slightly larger, say .501. In this case, we would expect that the proportion of heads in a sequence of tosses would tend to be slightly larger than 1/2. Thus, if in 99 tosses we observe more heads than tails (in the present case 56 heads versus 43 tails), we would actually be wiser to bet that a head will appear on the final toss than that a tail will appear. The reasoning is simple: If the coin *is* slightly biased rather than perfectly fair, the evidence accumulated thus far tends to indicate that a head is slightly more likely to appear than is a tail. If, in addition, we rid ourselves of the rather ridiculous notion that the coin has a memory, a questionable knowledge of statistical theory, and a compulsive desire to conform to what has been called the law of averages, we strengthen our belief that betting on heads on the final toss is slightly preferable to betting on tails.

So much for the law of averages!

Remark. There are many generalizations of the law of averages. One generalization, described in the fanciful imagery of a large number of monkeys steadily typing away in the jungle, shows that eventually one of them will type the complete works of Shakespeare (in chronological order without a single error!). Also, another monkey will type the complete works of Proust, and finally, a third monkey will type Milton's epic poem *Paradise Lost*. Results of this nature intrigue probabilists and boggle the sensitive minds of the literati.

Summary

1. The chance is high that the proportion of heads in a large sequence of tosses of a fair coin is close to .5. The greater the number of tosses, the smaller is the chance that the proportion of heads differs from the theoretical value .5·by more than any specified small amount. Roughly speaking, the observed proportion converges to the theoretical value .5 as the number of tosses gets very large.
2. Note that the statement in 1 above concerns the chance of future outcomes and does not state definitively that any specified outcome (for example, a head on the next toss) or set of outcomes will definitely occur.
3. An inanimate object such as a coin, card, or die does not have a memory (a computer memory notwithstanding).

Problems

1. Suppose a coin is very carefully fabricated so that *we have complete assurance that a head and a tail are equally likely; that is, each outcome has probability 1/2.* After 99 tosses, we have obtained a certain number of heads and a certain number of tails. In betting on the outcome of the next toss, should we take into account the number of

SMITH CORONA II

" . . . that this nation, under God, shall have a new birth of freedom–and that government of the people, by the people, for the people, shall not perish from the pz4$ya!"

    heads and the number of tails that have appeared thus far?
        Justify your answer.

2. Suppose instead, the probability of a head turning up can be estimated only by observing its relative frequency in a long sequence of trials. Suppose further that the outcome in 100 trials is 60 heads and 40 tails. On the next trial, would you bet on a head or on a tail (even-money bet)? Justify your answer.

3. Suppose we toss a *fair* coin:
    a. 100 times

b. 1,000 times

c. 10,000 times

Which of the three cases is most likely to yield a deviation of at most 5% from the ideal 50% heads? Explain.

4. A fair coin is tossed

a. 2 times

b. 4 times

What is the chance of getting heads *exactly* half the time in a, in b? Explain the apparent anomaly.

Hint: List all possible outcomes in a. What proportion of them corresponds to 50% heads? Now perform a similar listing in b and answer the same question.

5. A coin is tossed 100 times yielding 50 heads in succession followed by 50 tails in succession. What conclusions do you reach?

6. "It's the bottom of the ninth, folks, and Garvey has not reached base in his last 15 at-bats. He is due for a hit." Criticize this baseball announcer's statement in view of what you learned from this vignette. Is a sequence of at-bats in baseball completely analogous to a sequence of coin tosses? Explain.

7. Is there any merit in the following gambling scheme at red-black roulette? Wait until there is a sequence of seven straight black outcomes and then put your money on red on the next spin of the wheel. Explain your answer.

8. Describe two business situations where people incorrectly invoke the so-called law of averages.

9. An aspiring young actor has been rejected at his last 24 auditions. He goes to his next audition with very high hopes. Is he invoking a law of averages? Is this situation analogous to coin-tossing?[†]

10.* Lincoln E. Moses has stated: "There are no facts for the future." Yet he and other statisticians do attempt predictions (and hedge their bets by stating chances of errors). If a statistician (or anyone else, for that matter) cannot predict with certainty whether or not a specific outcome will actually occur, what is the value of statistics and probability?[‡]

[†]Although the actor is clearly misusing the law of averages, we cannot help but envy him. Statistically, he has every reason to be pessimistic concerning his forthcoming audition, yet because of his faulty understanding of the law of averages, he ends up very hopeful. Undoubtedly, he performs better because of his grossly mistaken conception of the law of averages.

Can it be that truth is a two-edged sword? "Where ignorance is bliss, t'is folly to be wise."

Late bulletin! The aspiring actor was accepted for the role and continued on to build a long and successful career in the movies. Currently he is president of the United States. His legislative program calls for Congress to pass the law of averages as he knows it.

[‡]Oy veh! Am I glad I don't have to answer this question! Coauthor.

Appendix I
Tables

Table A   10,000 Random Digits

07048	52841	54969	87057	30570	50494	29936	93967	10641	79871
09165	56926	17294	03803	31755	11321	33681	12997	17625	25954
35654	69761	83791	63371	28189	19944	04514	56533	89108	27861
79065	63956	39443	30373	55571	00919	15377	36851	28318	40846
27969	74368	77782	88616	06368	07345	00725	81221	78417	37992
47528	70548	25078	80729	27806	42877	80287	21759	61980	52447
65694	95760	64031	24046	77606	91163	51492	20958	18384	49840
24253	39427	80642	36718	92164	77732	69754	01291	53704	33054
34302	60309	27186	22418	59962	13934	67591	17476	21559	73437
76809	84341	74012	50947	83214	19967	44219	75929	13182	34858
85183	35958	04301	49628	91493	66103	65699	04241	82441	38112
27541	79187	99777	22894	83283	56218	86183	74497	21070	78935
74188	09083	54938	79920	27158	24864	31116	33173	43032	52000
13270	57457	30968	65978	67679	91216	47969	39204	46030	93954
89150	53922	40537	23169	46948	05519	72171	85417	31580	98102
49980	44551	99908	46115	92508	77184	44556	69725	42878	60298
26810	40280	15387	30976	15478	77703	34109	02682	52877	36755
35056	23942	42645	67063	44118	46433	83172	95689	60923	32769
09873	65959	77912	70059	07704	16015	57527	09818	84379	35903
40806	30051	54251	73489	47215	90651	90083	21019	63860	41369
21845	81166	51104	72709	31590	99908	26621	49106	43778	55985
44709	95448	39230	70674	93354	80297	42761	99321	34882	95090
88151	35769	99767	63152	29152	70618	34376	67767	35703	28539
66501	12509	02489	04199	95426	03612	24007	12422	47324	50917
45920	06860	25503	93400	61029	03446	39355	66271	18495	84619
63759	47371	90159	69524	42653	68042	80018	42376	71319	02912
33214	50932	52735	66609	59419	69197	50268	73814	61258	65414
14340	44938	42375	48605	26169	41468	16476	50944	64859	76614
91608	55471	83143	53851	94514	21088	51344	89729	34715	23153
74473	05046	57268	24109	09978	10073	82888	17984	07699	89371
54499	91382	49612	85206	65024	94592	24353	38026	01543	50969
13171	63352	31070	13018	57884	36913	30883	62299	84146	19225
33063	31754	35455	15015	97981	18736	82504	65167	04552	19340

Table A *(continued)*

34208	61668	67083	36341	49647	13068	99928	61218	08212	27580
05059	09789	15892	20722	52379	88091	78381	09934	76456	70883
18096	86346	63455	92066	20412	43674	55777	28935	68363	25856
07561	34298	57759	87823	27897	12446	38084	13074	67260	37091
91910	97709	82077	76323	09540	48015	31320	09323	88534	38591
14035	97000	94897	75025	34413	10546	68378	28675	96204	44218
19174	58058	05722	48974	18904	24912	81304	58058	50631	25260
12493	96101	17526	13480	64797	42804	54137	59591	30638	33095
40353	56887	77472	02433	51784	37153	88978	87668	48367	85359
02980	77236	70259	36742	10470	50806	12963	83632	91395	71483
23310	72762	32854	83077	53036	72509	85910	56463	33606	43427
73857	22920	51975	72110	20445	65218	78164	48669	14957	24727
69391	06705	55659	28097	72354	82968	40825	98902	91264	96344
80367	19422	25217	92269	69652	27706	86501	70724	16864	65994
05258	45220	94413	40009	72207	79225	29924	61988	73682	63159
39000	31088	31119	72578	73000	32187	48620	31945	75758	31928
72033	70197	50311	47216	31096	25958	47262	67170	25090	44381
46493	71042	92806	16432	34497	99449	01365	28984	36640	89928
39930	93644	37525	88139	00302	93078	22352	89653	16752	75115
98489	19613	25939	59612	33099	29559	52187	85900	34103	31727
72878	86707	02696	25351	26136	08235	80878	45622	20838	06726
28490	62215	03404	10869	16591	70388	19529	69219	86120	71637
59574	80831	46674	02545	15061	13384	70654	78727	47793	54985
23324	92901	83965	53753	36657	67516	45039	07390	20899	32860
58218	60308	54831	23277	85480	69527	51651	12257	36539	98208
84406	40193	27182	82958	49691	11604	73169	11771	17805	33811
03585	05930	75527	84663	23911	02274	78215	58819	64314	25472
21078	62340	07418	13633	38849	21372	48988	77649	09561	44412
47552	37141	77185	74903	84254	86686	82103	51134	66713	03139
86708	28130	22052	45370	42107	90635	02798	28293	54153	50804
90375	36261	95090	90629	92850	27207	65037	89965	31574	87846
91084	02196	37716	12405	57210	85320	30334	06192	98645	19091
97428	79349	84149	76871	61394	75622	74968	52750	13264	55986
57643	99604	04861	37328	41212	60514	45687	57103	81791	87003
95128	75062	00508	72538	09177	35460	43437	56059	60499	07650
55232	45375	29178	94847	65615	92710	81079	48516	88940	67210
08749	55174	73670	82005	02382	87132	08269	10952	57997	66775
03419	52186	22773	41770	88607	93271	95663	06338	65962	05971
03719	13680	94088	50660	64259	75300	85329	95332	45152	47308
70806	40772	81890	76601	34540	91153	93964	18518	50646	90946
19172	55360	50005	42851	67586	77752	27132	24426	57417	92315
43407	47311	39205	37142	76288	30518	57718	08655	15647	62512
19863	65928	18030	16615	75512	20825	78212	87973	59172	64126
20254	88859	62423	35905	95138	62997	47227	51690	23462	98853
46935	34335	93215	70317	14714	91641	86466	68850	01407	17679
46630	00089	23947	38048	53611	17604	64639	63131	09135	60391
99593	11334	81791	16630	59435	87916	32618	83890	04181	71787

Table A *(continued)*

49300	81579	34574	00383	86360	29364	35100	11053	45724	85718
19570	03330	48752	85675	47019	46927	64434	92896	69414	86072
25411	07910	38481	50352	00269	70512	12438	32809	98785	69722
78343	65348	95304	54222	16367	32527	06433	10633	35807	03056
97055	78602	92450	29758	53815	58496	83643	49702	64230	80747
99429	64179	43718	05323	27311	02442	03317	11237	58161	71887
68632	86640	03245	37708	30397	33409	73883	25890	55720	14151
30691	81408	09566	17355	35806	66351	00840	54760	44150	96337
13019	02030	74721	45540	08843	88278	97523	61196	15708	34335
24357	20612	82043	23009	69041	07394 ·	11717	74913	35418	35638
40449	71327	38807	27862	29425	72466	70936	30801	62593	98224
28819	59123	47445	31946	53122	09967	01556	62394	66602	56768
43359	90350	99065	46690	17786	70835	75482	15215	53008	13219
59788	73061	48300	74982	18386	75577	47373	33601	63397	17236
33719	04524	45392	52791	50619	79443	29496	05349	34387	93762
91614	63890	96824	09615	56130	66707	59311	16204	22324	60729
80574	72453	53819	09649	68208	89625	22294	79244	14553	65118
44635	06487	68348	49446	02089	09275	95745	92351	24232	01109
19476	96891	25812	53629	92557	28157	80217	66529	10579	32145
06317	22015	14079	17033	72820	15777	49207	96350	74103	12570
69743	90929	18548	30408	83108	52135	16381	11767	30376	85231
98532	40050	73954	10216	39541	14588	54213	71954	01305	39748
33464	59848	82734	41344	75572	50951	88917	13174	29666	35957
26496	09868	97384	83052	51586	22242	97808	32239	18567	10565
35864	49715	61236	17423	20491	74166	55708	70644	85201	65104
50580	98106	23850	23505	91224	10852	41394	10373	10915	27128
85954	22867	14199	18348	58351	39251	07604	48221	37475	23727
25604	59590	98866	56998	91046	84941	96552	34628	35850	05788
06089	39461	37176	96378	01667	74786	36020	00088	49849	69655
45206	32269	54176	37929	32742	58128	41221	30876	79157	76241
72507	98266	96505	45160	70248	42029	38268	85295	62172	43116
04704	21320	89204	54174	16390	32542	11123	42694	54949	07431
67118	42511	73852	92587	95400	15760	52025	92659	17652	55595
69795	29615	64876	41851	82171	50300	91905	78569	26794	76533
54816	80034	47404	88673	03364	14991	33284	61527	85134	38008
99460	43676	30662	62295	39942	81533	94007	08029	78865	05481
58871	01188	28430	96911	58375	79509	08923	78391	69726	91828
50987	52808	43715	79437	16502	67263	04601	75491	19044	85303
16802	91568	80308	65430	32311	71683	35540	09610	88680	43306
57366	06681	76872	42158	63785	75819	80098	83313	97434	13435
05518	14389	22545	05839	22325	50125	67380	91125	69686	27203
74242	80437	21199	79605	98368	33730	04054	25842	47759	03325
21561	60139	15271	76752	09849	19036	51270	32119	47665	92296
87324	65972	14132	43464	22452	20480	65595	60332	17830	74709
25627	38413	07639	29043	67131	59025	44199	12072	64067	49562
43801	39932	45979	47695	16029	36798	68562	60394	24422	76913
21093	91670	34623	82583	94811	40565	20059	44620	18884	15973

Table A *(continued)*

93734	01115	93668	98210	74351	74511	59558	13946	84653	26619
78046	36556	52983	63314	63428	62772	81475	82588	04863	67553
13856	90217	62338	07427	88600	51436	85819	35322	98239	04479
60629	34594	94757	23990	53285	89621	26727	89836	00132	73218
40534	96125	89065	17245	23000	40886	13436	70122	39038	47038
08123	31849	50177	27095	87986	51022	33837	87284	86968	03021
88032	02832	87185	31683	41726	49981	03380	78201	49734	90899
46721	30942	83460	49679	45815	13050	33796	94097	88317	90395
75684	64787	54889	33308	77806	24991	12767	68376	44835	58774
41644	78813	31303	00101	60401	48080	43670	09732	42827	43263
08533	40604	72071	19546	91230	41711	08097	88032	35547	88263
62267	83359	76489	88492	67103	53240	38754	22448	79734	86600
45188	40211	31521	67605	64452	61671	95814	14328	15551	66260
83708	97234	39653	13413	58656	99094	37725	17625	97754	38730
47682	11510	39769	96286	45477	67789	53056	90585	98860	10947
47274	24009	09586	79738	26668	98156	43251	70613	18510	27580
47764	93802	59122	36233	74756	14387	33719	24862	65268	23591
65398	95217	44823	58024	02229	39160	99509	23795	23549	23725
90200	03072	16926	00779	74304	70657	99731	95659	39473	00534
76116	41731	93522	40654	13201	76237	51760	93859	16655	42936
10500	82529	70573	60023	36437	91822	19897	83175	92020	40368
73129	51104	11001	72445	35826	48508	58795	83000	04848	91628
12789	39825	58290	28556	08074	47142	45621	26831	69227	64547
49431	48619	20684	98122	05325	59878	08651	15627	11928	71035
83861	41642	26414	03780	59875	05511	84753	81327	46369	96872
12191	41142	70423	10673	19083	53688	57121	16186	16009	64075
68794	14525	09448	63406	57620	03452	63541	80623	23772	32250
05621	47985	05287	07703	12624	33067	68297	40827	41467	31132
49624	17243	86462	90171	97999	65911	59913	05648	26039	01568
41909	87188	24948	94285	85120	45938	05733	02121	10397	31038
05343	98714	36298	16914	42719	55448	96582	26976	02142	97862
98512	64106	76781	27644	07210	92564	91955	40493	92088	16851
98960	08043	80681	83743	67085	09666	31387	13099	62719	33629
34583	79520	65947	58060	93289	53229	27319	34884	38461	62561
89098	26561	26153	44042	99546	25669	73824	02044	77748	83543
63754	15419	51882	97195	17248	95121	42889	06868	34246	32235
84061	02526	86747	54370	07161	11729	15078	27129	93875	98234
69684	40905	06480	49353	68698	37329	10905	65737	20836	25374
93257	79590	52694	53238	08085	13045	99055	65876	78512	98754
55160	45972	83651	12782	11737	85403	09291	51762	52977	17206
29250	69956	99529	44047	06044	98492	27429	35026	40592	62936
49305	68764	83680	34372	11988	79542	20655	54004	03527	71832
18033	69707	99173	67768	80685	51423	66920	70093	75441	34691
01160	33167	78354	56227	73914	53317	90068	19860	16457	12128
76715	76806	12275	14069	35637	57161	61396	72119	29033	29258
73247	65102	51278	73727	95424	64525	80920	30584	26087	31038
66151	49015	44649	75095	00621	78527	34183	08291	93119	43512
92260	24470	53500	79066	65867	03494	51468	05934	58966	84999

Table A *(continued)*

43172	73110	64715	39562	87842	64493	62084	58285	74974	25984
87821	90155	12033	22806	62168	30047	93345	50910	74420	86666
09578	25920	98170	78504	64802	23848	29439	08808	91012	69003
63977	42111	34052	56483	47962	68022	90686	18597	78176	65511
77832	56425	90832	52415	49239	30042	37997	46134	72535	91305
12080	25428	24391	64774	67400	51673	59666	16347	34117	92104
10650	32743	27699	00018	40133	60987	79371	39442	07547	30622
93641	29860	74783	35944	75404	70978	66492	39981	99811	17672
61701	16635	42118	93756	98771	80811	10757	56592	81283	52868
91414	89835	20333	33143	16123	41923	59084	62832	99609	56362
03146	43473	54396	25962	36812	66127	56521	57405	04824	86397
61367	85905	03423	44865	26837	90048	92947	92328	06614	94099
60945	96129	86198	04027	48749	29877	76559	25002	33225	40023
42729	66044	86315	34817	39207	70910	77802	51450	91379	08861
23261	00945	46969	38421	27584	72368	20636	15145	13125	11115
49094	61963	33657	89000	26723	29978	41734	52489	16493	85672
48707	25054	71655	97204	88853	55973	60684	77774	12868	98186
02307	53148	89028	98443	87467	64066	59429	51735	82919	37113
17383	08466	97122	26713	15628	13012	39423	77178	08197	34378
77241	29608	19332	68534	10088	91994	63254	44291	17475	52594
64800	14797	65166	06006	18060	45007	58955	29955	64573	93626
54237	22540	76604	34289	90463	78043	55982	92762	98965	16481
89803	42065	17562	63450	19802	76595	78021	75691	76104	45987
11749	34925	34209	48598	72301	09922	53339	80262	59603	09099
82461	58452	12385	17475	50531	30644	94395	87144	12384	07328

Table B   Binomial Coefficients

Example: $\binom{5}{2} = 10$

Some interpretations:

1. A sample is a selection of some of the objects from a group of objects. The number of distinct samples of size 2 that may be selected from 5 different objects is 10. For example, from the group {a, b, c, d, e} we may select the following samples of size 2: (a, b), (a, c), (a, d), (a, e), (b, c), (b, d), (b, e), (c, d), (c, e), (d, e).

2. There are 10 possible arrangements of 2 females and 3 males: FFMMM, FMFMM, FMMFM, FMMMF, MFFMM, MFMFM, MFMMF, MMFFM, MMFMF, MMMFF.

Size of Group	Size of Sample							
	2	3	4	5	6	7	8	9
2	1							
3	3	1						
4	6	4	1					
5	10	10	5	1				
6	15	20	15	6	1			
7	21	35	35	21	7	1		
8	28	56	70	56	28	8	1	
9	36	84	126	126	84	36	9	1
10	45	120	210	252	210	120	45	10
11	55	165	330	462	462	330	165	55
12	66	220	495	792	924	792	495	220
13	78	286	715	1,287	1,716	1,716	1,287	715
14	91	364	1,001	2,002	3,003	3,432	3,003	2,002
15	105	455	1,365	3,003	5,005	6,435	6,435	5,005
16	120	560	1,820	4,368	8,008	11,440	12,870	11,440
17	136	680	2,380	6,188	12,376	19,448	24,310	24,310
18	153	816	3,060	8,568	18,564	31,824	43,758	48,620
19	171	969	3,876	11,628	27,132	50,388	75,582	92,378
20	190	1,140	4,845	15,504	38,760	77,520	125,970	167,960
21	210	1,330	5,985	20,349	54,264	116,280	203,490	293,930
22	231	1,540	7,315	26,334	74,613	170,544	319,770	497,420
23	253	1,771	8,855	33,649	100,947	245,157	490,314	817,190
24	276	2,024	10,626	42,504	134,596	346,104	735,471	1,307,504
25	300	2,300	12,650	53,130	177,100	480,700	1,081,575	2,042,975
26	325	2,600	14,950	65,780	230,230	657,800	1,562,275	3,124,550
27	351	2,925	17,550	80,730	296,010	888,030	2,220,075	4,686,825
28	378	3,276	20,475	98,280	376,740	1,184,040	3,108,105	6,906,900
29	406	3,654	23,751	118,755	475,020	1,560,780	4,292,145	10,015,005
30	435	4,060	27,405	142,506	593,775	2,035,800	5,852,925	14,307,150

Size of Group	Size of Sample					
	10	11	12	13	14	15
10	1					
11	11	1				
12	66	12	1			
13	286	78	13	1		
14	1,001	364	91	14	1	
15	3,003	1,365	455	105	15	1
16	8,008	4,368	1,820	560	120	16
17	19,448	12,376	6,188	2,380	680	136
18	43,758	31,824	18,564	8,568	3,060	816
19	92,378	75,582	50,388	27,132	11,628	3,876
20	184,756	167,960	125,970	77,520	38,760	15,504
21	352,716	352,716	293,930	203,490	116,280	54,264
22	646,646	705,432	646,646	497,420	319,770	170,544
23	1,144,066	1,352,078	1,352,078	1,144,066	817,190	490,314
24	1,961,256	2,496,144	2,704,156	2,496,144	1,961,256	1,307,504
25	3,268,760	4,457,400	5,200,300	5,200,300	4,457,400	3,268,760
26	5,311,735	7,726,160	9,657,700	10,400,600	9,657,700	7,726,160
27	8,436,285	13,037,895	17,383,860	20,058,300	20,058,300	17,383,860
28	13,123,110	21,474,180	30,421,755	37,442,160	40,116,600	37,442,160
29	20,030,010	34,597,290	51,895,935	67,863,915	77,558,760	77,558,760
30	30,045,015	54,627,300	86,493,225	119,759,850	145,422,675	155,117,520

Appendix II

Solutions to Selected Problems

Vignette 1

1. (a1) 18/38. (a2) 20/38.

(b1) Since red and black are equivalent in the sense that each has a chance of 18/38 of occurring, and since the probability structure of each spin is the same, it suffices to consider a single bet of $1 on black. The expected value of a random quantity which can take on a finite set of values is obtained by taking each possible value of the quantity, multiplying the possible value by its probability, and summing all the products of the form "possible value" × "probability of that value," there being one product for each possible value. For the random quantity "gain to the player" the possible values and probabilities are

Possible Value	Probability of that Value	Product
1	$\dfrac{18}{38}$	$1 \times \dfrac{18}{38}$
-1	$\dfrac{20}{38}$	$-1 \times \dfrac{20}{38}$

The expected gain of the player is obtained by adding the products:

$$1 \times \frac{18}{38} + \left(-1 \times \frac{20}{38} \right) = -\frac{2}{38},$$

or equivalently the expected loss (or long-run percentage loss) of the player is $(2/38) \times 100 = 5.3\%$.

(b2) From (b1) the long-run percentage gain of the house is seen to be 5.3%.

(c) No, a single bet is bad enough. Combining bets compounds the dilemma (gives the house more opportunities to gain from their edge). Since the wheel has no memory, strategies which take into account the past do not help.

(d) He should use bold play and bet the $100 on red (or on black, it doesn't matter). His chance of reaching his goal of winning $100 is 18/38. This system is better than competing systems because (roughly speaking) it doesn't let the house's steady advantage over the bettor continually operate or accrue.

(e) To make the bet a fair bet the expected gain to the player should be zero. Then the expected gain to the house will also be zero. Thus the "amount house should pay" (in addition to returning the original $1 bet), if the player bets $1 on red and red appears, should satisfy the following equation:

"amount house should pay" $\times$ "probability player wins"
    $-$ "amount player pays" $\times$ "probability player loses" $= 0$.

This reduces to

$$\text{"amount house should pay"} \times \frac{18}{38} - 1 \times \frac{20}{38} = 0,$$

or

$$\text{"amount house should pay"} = \frac{20/38}{18/38} = \frac{20}{18} = \$1\frac{2}{18}.$$

The house should pay according to 20 to 18 odds rather than the 1 to 1 unfair payoff that it uses.

4. There cannot be a strategy that gives a 1/10 (or greater) chance of turning $1,000 into $10,000. For if there were such a strategy, the expected total (including the initial $1,000) from such a strategy would be

$$10,000 \times \text{"some fraction that is greater than or equal to} \frac{1}{10}\text{",}$$

and the above product would then be some number greater than or equal to 1,000. That would say there is a strategy which yields a positive expected gain. But we know this is impossible because each individual bet is unfavorable to the player and, thus, all combinations of bets must be unfavorable.

9. Some gamblers simply enjoy the actual act of betting and hanging around the roulette table for the atmosphere, and so on. Such gamblers may be willing to let their money dwindle at a slow rate. Many compulsive gamblers are willing to lose money just for the fun of it and like to take their time doing it.

Vignette 2

1. The six possible permutations are 123, 132, 213, 231, 312, 321.

The strategy "pass 0 draws, then accept the first candidate thereafter" (that is, pick the first draw) wins for 312 and 321 and thus has probability 2/6 of winning.

The strategy "pass 1 draw, then accept the first candidate thereafter, if any" wins for 132, 213, 231 and thus has probability 3/6 of winning.

The strategy "pass 2 draws, then accept the first candidate thereafter" (that is, pick the third draw) wins for 123 and 213 and thus has probability 2/6 of winning.

Thus the best strategy is "pass 1 draw, then accept the first candidate thereafter, if any." The chance of winning with this strategy is .5.

6. Refer to Table 2.1 of Vignette 2 and note that the number 4 appears in the first position in 6 of the 24 permutations, in the second position in 6 of the permutations, in the third position in 6 of the permutations, and in the fourth position in 6 of the permutations. Thus, for each of the strategies I, II, III, IV the chance of selecting the largest dowry is 1/4.

8. The opponent cannot reduce the chance that the player will win to a value less than 1/4 because the player can always choose the position at random.

Vignette 3

1. The left-hand side of inequality (2) is .6, the right-hand side is $.8 \times .7 = .56$, and thus the two-point attempt is preferred.

2. The left-hand side of inequality (2) is .7, the right-hand side is $.9 \times$ [(utility for tying)/(utility for winning)]. Thus, for indifference we must have

$$.7 = .9 \times \frac{\text{(utility for tying)}}{\text{(utility for winning)}},$$

or equivalently

$$\frac{\text{utility for tying}}{\text{utility for winning}} = \frac{7}{9}.$$

8. Yes. Practice makes (almost) perfect.

Vignette 4

1. The chance that all three engines function is $.99 \times .99 \times .99$ $= .9703$. The chance that engines one and two function but engine three does not is $.99 \times .99 \times .01 = .0098$. Similarly the chance that engines one and three function but engine two does not is .0098, and the chance that engines two and three function but engine one does not is .0098. Thus the reliability of the airplane is

$$.9703 + .0098 + .0098 + .0098 = .9997.$$

2. (a) We assume excessive radiation is present and calculate the chance that the monitor fails to detect it. This is equal to the

"Chance that none of the detection devices claims excessive radiation"
+ "Chance that exactly one of the detection devices claim excessive radiation."

The chance that none of the detection devices claims excessive radiation (when excessive radiation is present) is $.01 \times .01 \times .01 \times .01 = .00000001$. The chance that exactly one claims excessive radiation (when excessive radiation is present) is

$$.99 \times .01 \times .01 \times .01 \ + \ .01 \times .99 \times .01 \times .01 \ + \ .01 \times .01 \times .99 \times .01 \ + \ .01 \times .01 \times .01 \times .99$$

↑	↑	↑	↑
Chance that device one claims excessive radiation and other three devices do not	Chance that device two claims excessive radiation and other three devices do not	Chance that device three claims excessive radiation and other three devices do not	Chance that device four claims excessive radiation and other three devices do not

$$= .00000099 + .00000099 + .00000099 + .00000099 = .00000396.$$

Thus, the chance that the monitor fails to detect excessive radiation is $.00000001 + .00000396 = .00000397$.

(b) We assume excessive radiation is not present and calculate the chance that at least two out of the four detection devices claim excessive radiation is present. This chance is equal to

"Chance that exactly two claim excessive radiation"
+ "Chance that exactly three claim excessive radiation"
+ "Chance that all four claim excessive radiation."

By calculations similar to those in part (a) these chances are computed as follows:

Chance that exactly two claim excessive radiation

$$= \quad .01\times.01\times.99\times.99 \quad + \quad .01\times.99\times.01\times.99 \quad + \quad .01\times.99\times.99\times.01$$

↑   ↑   ↑

Chance that devices   Chance that devices   Chance that devices
one and two give false   one and three give   one and four give false
alarms and three and   false alarms and two   alarms and two and
four do not   and four do not   three do not

$$+ \quad .99\times.01\times.01\times.99 \quad + \quad .99\times.01\times.99\times.01 \quad + \quad .99\times.99\times.01\times.01$$

↑   ↑   ↑

Chance that devices   Chance that devices   Chance that devices
two and three give   two and four give false   three and four give false
false alarms and one   alarms and one and   alarms and one and
and four do not   three do not   two do not

$$= 6 \times .99 \times .01 \times .01 \times .99 = .000588.$$

Chance that exactly three claim excessive radiation

$$= .01 \times .01 \times .01 \times .99 + .01 \times .01 \times .99 \times .01$$
$$+ .01 \times .99 \times .01 \times .01 + .99 \times .01 \times .01 \times .01$$

$$= 4 \times .01 \times .01 \times .01 \times .99 = .00000396.$$

Chance that all four claim excessive radiation

$$= .01 \times .01 \times .01 \times .01 = .00000001.$$

Adding, we obtain the desired probability to be

$$.000588 + .00000396 + .00000001 = .000592.$$

5. $10 \times 10 \times (1 - .9997) = .03.$

Vignette 5

1.   With 100% inspection of a product for which inspection results in destruction of the product, 100% inspection leaves the manufacturer with no product to sell. Obviously, statistical sampling plans are to be preferred.

2.   (a) Approximately .90. (b) Approximately .05. (c) Approximately .078.

6.   One possible explanation: 100% inspection of a large lot is tedious and boring and the inspectors may become careless and make errors concerning the quality of the product.

Vignette 6

4. (a)

Sample Number	Sample Mean
30	.8318
31	.8358
32	.8426
33	.8294
34	.8286
35	.8294
36	.8300
37	.8298
38	.8236
39	.8146

(c) The sample means corresponding to sample numbers 32 and 39.

5. An alternative measure of variability is the **sample variance**. The sample variance is computed as follows:

a. Compute the sample mean.
b. Subtract the sample mean from each observation to obtain, for each observation, a difference of the form
   observation − sample mean.
c. Square each difference.
d. Sum the squared differences.
e. Divide this sum by the quantity "sample size − 1," to obtain the sample variance.

Example: Consider a sample consisting of the five values 1, 6, 7, 8, 12. The sample mean is $(1 + 6 + 7 + 8 + 12)/5 = 6.8$.

Difference Between Observation and Sample Mean	Difference Squared
$1 - 6.8 = -5.8$	$(-5.8)^2 = 33.64$
$6 - 6.8 = -\ .8$	.64
$7 - 6.8 =\quad .2$	.04
$8 - 6.8 =\quad 1.2$	1.44
$12 - 6.8 =\quad 5.2$	27.04
	Sum: 62.8

$$\text{Sample variance} = \frac{62.8}{4} = 15.7.$$

The **sample standard deviation**, another measure of variability, is defined to be the square root of the sample variance. In our example, the sample standard deviation is $\sqrt{15.7} = 3.96$.

8.   Suppose, for example, that there are two different work crews, one for Monday, Wednesday, Friday, the second for Tuesday, Thursday. If the second crew commits errors that cause the process to be out of control, such errors will not be detected by the charts constructed for Monday, Wednesday, Friday.

Vignette 7

3.   To solve this problem it is helpful to know some elementary rules of probability. We list some of these rules.

I.   The probability of an event is a number between 0 and 1 indicating how likely the event is to occur.

II.   **Complementary Rule for Probability.**

"Probability that an event will not occur"
     $= 1 -$ "Probability that it will occur."

III.   **Addition Rule for Mutually Exclusive Events.**   Events A,B are said to be mutually exclusive if they cannot occur simultaneously. If events A,B are mutually exclusive, then

"Probability that event A or event B will occur"
     $=$ "Probability that A will occur" + "Probability that B will occur."

This rule is extended to more events in the obvious way. For example, for the case of three events A,B,C (say) which are mutually exclusive (that is, no two or more can occur simultaneously),

"Probability that event A or event B or event C will occur"
     $=$ "Probability that A will occur" + "Probability that B will occur"
     + "Probability that C will occur."

IV.   **Multiplication Rule for Independent Events.**   Events A,B are said to be independent if the chance of A occurring is unaffected by knowledge of whether or not B occurred. Now suppose that A,B are independent. Then

"Probability that both events will occur"
= "Probability that A will occur" $\times$ "Probability that B will occur."

This rule is extended to more events in an obvious way. We say three events A,B,C (say) are mutually independent if, roughly speaking, the chance of each event is unaffected by knowledge of the outcomes (occur, not occur) of the other two events. Now suppose that events A,B,C are mutually independent. Then

"Probability that all three events will occur"
= "Probability that A will occur" $\times$ "Probability that B will occur"
$\times$ "Probability that C will occur."

We now return to the systems given in Problem 3. The reliability of the system shown in (a) is equal to the probability that all three amplifiers function. By Rule IV, this probability is equal to

"Probability that amp. 1 functions"$\times$"Probability that amp. 2 functions" $\times$ "Probability that amp. 3 functions"
$= .95 \times .95 \times .95 = .857.$

To determine the reliability of the system shown in (b), let A be the event that "all three of amplifiers 1, 2, 3 function" and let B be the event that "all three of amplifiers 1', 2', 3' function." Then by the calculation we have already made to compute the reliability of system (a),

"Probability of A" = "Probability that subsystem (1, 2, 3) functions"
$= .95 \times .95 \times .95 = .857,$
"Probability of B" = "Probability that subsystem (1', 2', 3') functions"
$= .95 \times .95 \times .95 = .857.$

Thus, by Rule II, the probability that subsystem (1, 2, 3) does not function is $1 - .857 = .143$; similarly, the probability that subsystem (1', 2', 3') does not function is $1 - .857 = .143$, and by Rule IV we find

"Probability that both subsystems (1, 2, 3) and (1', 2', 3') do not function"
$= .143 \times .143 = .020.$

Now the probability we wish to compute is:

"Reliability of system (b)"
= "Probability that at least one of the subsystems (1, 2, 3), (1', 2', 3') functions."

Applying Rule I we obtain

"Probability that at least one of the subsystems (1, 2, 3), (1', 2', 3') functions

  $= 1 - $ "Probability that both (1,2,3) and (1', 2', 3') do not function"

  $= 1 - .020 = .980.$

To determine the reliability of system (c), let A be the event that "at least one of amp. 1, amp. 1' functions," let B be the event that "at least one of amp. 2, amp. 2' functions," and let C be the event that "at least one of amp. 3, amp. 3' functions." Then

"Probability of A"

  $= 1 - $ "Probability that both amp. 1 and amp. 1' fail"

  $= 1 - (.05)^2 = .9975,$

"Probability of B" $= .9975,$

"Probability of C" $= .9975.$

The reliability of system (c) is equal to

"Probability that all three events A, B, C occur"

which by Rule IV is $.9975 \times .9975 \times .9975 = .993.$

Comparing systems (a) and (b) we see that redundancy increases reliability, and comparing systems (b) and (c) we see that redundancy at the component level yields greater reliability than redundancy at the system level (all other things being equal).

6. A major league baseball team assigns five or six pitchers to its bullpen for each game. These pitchers are there to be ready to relieve the starting pitcher if he does not pitch well or if he needs to be taken out of the game for other reasons (injury, pinchhitter, and so on).

8. Consider a fluorescent light bulb comprised of two tubes. If both tubes don't function, there is no light. We could call this state "0." If one of the tubes functions and the other does not, we have partial light. We could call this state "1." If both tubes function, we have full light and we could call this state "2."

Vignette 8

1. The system is capable of producing music if:

Event (1) at least one of the tuner, changer functions,
and Event (2) the amplifier functions,
and Event (3) at least one of the two speakers functions.

The probability of Event (1) can be obtained by calculating the probability that "both the tuner and changer fail" and subtracting that probability from 1. (Here we are using the **complementary rule of probability**. Namely, "Probability that Event A will occur" = 1 − "Probability that A will not occur.") Thus,

"Probability of Event (1)"
= 1 − "Probability that both the tuner and changer fail"
= 1 − (.01 × .02) = 1 − .0002 = .9998.
"Probability of Event (2)" = .995,
"Probability of Event (3)" = 1 − (.02 × .02) = 1 − .0004 = .9996.

Multiplying together the probabilities of events (1), (2), and (3) yields .9998 × .995 × .9996 = .9944, representing the probability that the system is capable of producing music.

2. (a)

Subsystem	Reliability	Unreliability
Computer	.97	.03
Generator 1	.99	.01
Generator 2	.99	.01

The system functions if:

Event (1) the computer functions,
and Event (2) at least one of the two generators functions.

"Probability of Event (1)" = .97.
"Probability of Event (2)" = 1 − "Probability that both generators fail"
= 1 − (.01 × .01) = 1 − .0001 = .9999.

Multiplying together the probabilities of Events (1) and (2) yields .97 × .9999 = .9699, representing the reliability of the system.

(b)

Subsystem	Reliability	Unreliability
Computer	.985	.015
Generator 1	.995	.005
Generator 2	.995	.005

Defining Event (1) and Event (2) as in part (a),

"Probability of Event (1)" = .985,
"Probability of Event (2)" = 1 − (.005 × .005) = .999975,

and the new, improved system reliability is $.985 \times .999975 = .9850$.

6. In a system consisting of subsystems arranged in series, increasing the number of subsystems *decreases* the reliability of the system. Suppose that you put one additional subsystem in series with the original system. Then

"Reliability of the new system"

↑

term 1

="Reliability of original system"×"Reliability of additonal subsystem."

↑   ↑

term 2   term 3

Since term 3 is less than or equal to 1, term 1 must be less than or equal to term 2.

Vignette 9

2. Under criterion 3 there may be more than one "best" team. For example, suppose there are 10 teams in the league. We could have:

"Probability that team 1 beats team 6 in a typical game" = .6,
"Probability that team 1 beats team 7 in a typical game" = .6,
"Probability that team 1 beats team 8 in a typical game" = .6,
"Probability that team 1 beats team 9 in a typical game" = .6,
"Probability that team 1 beats team 10 in a typical game" = .6,

and

"Probability that team 2 beats team 6 in a typical game" = .6,
"Probability that team 2 beats team 7 in a typical game" = .6,
"Probability that team 2 beats team 8 in a typical game" = .6,
"Probability that team 2 beats team 9 in a typical game" = .6,
"Probability that team 2 beats team 10 in a typical game" = .6.

Then according to criterion 3, both team 1 and team 2 are best. Furthermore, it is possible that other teams are best also. Obviously, 3 is a very weak criterion for being best.

4. One possibility: Say a team is best if against *each* other team in the league, it wins more than half of the games played. An advantage of this approach is that at most one team could be the best team in the league. A serious disadvantage is that there may be no best team. For example using this method there would not have been a best team in the National League or in the American League in the 1965 season.

5. In 18 games during 1965, Phildelphia defeated Milwaukee 12 times and suffered 6 losses, for a winning frequency of 2/3. Thus we would estimate the chance that Philadelphia wins two additional games from Milwaukee as $(2/3) \times (2/3) = 4/9$. By multiplying the two probabilities together we are assuming that the two games are independent. That is, we are assuming that the chance that Philadelphia (and thus the chance that Milwaukee) wins the second additional game is unaffected by the outcome of the first additional game. This assumption may be unrealistic.

Vignette 10

3. The professor's conclusions are tainted by the 35% nonresponse. The professor needs to do a follow-up on the nonresponders.

7. It should be a population of women who are using the device because they do not want to get pregnant at that time. Population parameters of interest are probabilities concerning time to pregnancy. For example, we would be interested in the probabilities that a woman using the device would not get pregnant within 2 years, within 3 years, and so on.

10. Suppose you are comparing two drugs designed to reduce pain due to arthritis. You may receive early convincing evidence that one drug has deleterious side effects and thus want to discard that drug immediately.

Vignette 11

3. (a) Children. (b) Adults. (c) No. The company would do better to arrange taste tests with groups of children watching a circus, baseball game, and so on.

5. One type of nonresponse occurs when the individual is not at home. This problem can be attacked by "call-backs." Sampling theorists have models to analyze the effects of call-backs. A standard method is to specify a minimum number of call-backs before the person is abandoned as one who can't be contacted.

Sometimes nonresponse occurs because questions are difficult. This can be alleviated by careful wording of the questions and pretesting of the questionnaire.

Sometimes nonresponse occurs because questions are sensitive and put people on the spot (for example, "Do you smoke marijuana?"). Statisticians have devised clever **randomized response techniques** which allow a person to answer in a manner that does not seriously jeopardize his privacy.[†]

[†] For discussions and ways of treating nonresponse, see the textbook *Sampling Techniques*, third edition, by William G. Cochran, Wiley, New York, 1977.

8. The fact that you have just received free samples naturally increases the chances that you will name that brand as your favorite and as the one you are now using. The poll is biased and advertising based on it would be unethical.

Vignette 12

5. Pick a starting point at random. Then read 100 consecutive one-digit numbers. Each time you encounter a 0, 1, 2, 3, or 4 call it "heads"; each time you encounter a 5, 6, 7, 8, or 9 call it "tails."

8. Yes and yes. The properties of equal likelihood and mutual independence are satisfied if you read from right to left or if you read vertically; they do not depend on the direction in which you read.

9. An extreme case is the following. The table has 20 rows, 50 digits in each row. All the digits in the first two rows are "0," all the digits in the next two rows are "1," and so forth, so that all the digits in the last two rows are "9."

Vignette 13

1. (a)

Sample	Max. Est.	Adj. Est.	Sample	Max. Est.	Adj. Est.
1, 2, 3, 4	4	4	1, 4, 5, 7	7	31/4
1, 2, 3, 5	5	21/4	1, 4, 6, 7	7	31/4
1, 2, 3, 6	6	26/4	1, 5, 6, 7	7	31/4
1, 2, 3, 7	7	31/4	2, 3, 4, 5	5	21/4
1, 2, 4, 5	5	21/4	2, 3, 4, 6	6	26/4
1, 2, 4, 6	6	26/4	2, 3, 4, 7	7	31/4
1, 2, 4, 7	7	31/4	2, 3, 5, 6	6	26/4
1, 2, 5, 6	6	26/4	2, 3, 5, 7	7	31/4
1, 2, 5, 7	7	31/4	2, 3, 6, 7	7	31/4
1, 2, 6, 7	7	31/4	2, 4, 5, 6	6	26/4
1, 3, 4, 5	5	21/4	2, 4, 5, 7	7	31/4
1, 3, 4, 6	6	26/4	2, 4, 6, 7	7	31/4
1, 3, 4, 7	7	31/4	2, 5, 6, 7	7	31/4
1, 3, 5, 6	6	26/4	3, 4, 5, 6	6	26/4
1, 3, 5, 7	7	31/4	3, 4, 5, 7	7	31/4
1, 3, 6, 7	7	31/4	3, 4, 6, 7	7	31/4
1, 4, 5, 6	6	26/4	3, 5, 6, 7	7	31/4
			4, 5, 6, 7	7	31/4

(b) There are $\binom{7}{3} = 35$ possible samples, and they are equally likely. By adding values in the adjusted estimator columns in part (a) and dividing by 35, the average value of the adjusted estimator is found to be

$$\frac{4 + (21/4) + \cdots + (31/4)}{35} = \frac{(980/4)}{35} = 7.$$

2. By adding the values in the maximum estimator columns in part (a), and dividing by 35, the average value of the maximum estimator is found to be

$$\frac{4 + 5 + \cdots + 7}{35} = \frac{32}{5} = 6.4.$$

4. Adjusted estimator = [(maximum estimator) $\times$ (4/3)] $-$ 1. There are $\binom{7}{4} = 35$ possible samples, and they are equally likely. They are listed below, along with the corresponding values of the maximum estimator and adjusted estimator.

Sample	Max. Est.	Adj. Est.	Sample	Max. Est.	Adj. Est.
1, 2, 3	3	3	2, 3, 6	6	7
1, 2, 4	4	13/3	2, 3, 7	7	25/3
1, 2, 5	5	17/3	2, 4, 5	5	17/3
1, 2, 6	6	7	2, 4, 6	6	7
1, 2, 7	7	25/3	2, 4, 7	7	25/3
1, 3, 4	4	13/3	2, 5, 6	6	7
1, 3, 5	5	17/3	2, 5, 7	7	25/3
1, 3, 6	6	7	2, 6, 7	7	25/3
1, 3, 7	7	25/3	3, 4, 5	5	17/3
1, 4, 5	5	17/3	3, 4, 6	6	7
1, 4, 6	6	7	3, 4, 7	7	25/3
1, 4, 7	7	25/3	3, 5, 6	6	7
1, 5, 6	6	7	3, 5, 7	7	25/3
1, 5, 7	7	25/3	3, 6, 7	7	25/3
1, 6, 7	7	25/3	4, 5, 6	6	7
2, 3, 4	4	13/3	4, 5, 7	7	25/3
2, 3, 5	5	17/3	4, 6, 7	7	25/3
			5, 6, 7	7	25/3

(a) Since all 35 possible samples are equally likely, and since the adjusted estimator overestimates the true total 7 in 15 samples (the 15 samples where the adjusted estimator is 25/3), the chance that the adjusted estimator will overestimate the true total is 15/35. (b) 10/35 (c) 10/35 (d) 15/35

Vignette 14

5. The proportion of tagged fish in the recaptured sample is equal to

$$\frac{\text{number of tagged fish in recaptured sample}}{\text{size of recaptured sample}}.$$

The proportion of tagged fish in the population is equal to

$$\frac{\text{number of tagged fish in population}}{\text{size of population}}.$$

Equating these two proportions is equivalent to saying that you expect the proportion in the recaptured sample to be a good estimate of the proportion in the population. By equating the two proportions, and solving for the quantity "size of population," you obtain the estimate displayed in Problem 5 of Vignette 14.

7. Then the estimate of the size of the population is infinite—because the denominator of the term

$$\frac{\text{size of recaptured sample}}{\text{number of tagged fish in recaptured sample}}$$

is zero. But we know that the number of fish in the lake is some finite value.

8. You may have to wait a long time until the specified number of tagged fish are recaptured.

Vignette 15

1. (a) The sum of the proposed prices is $10 + 20 \times (1 + .7) = 44$. Thus proposal 1 should be chosen with probability 10/44 and proposal 2 should be chosen with probability 34/44.

(b) If the first proposal is the one selected in the sample, the government finds there is no padding factor and pays the contractor's prices, namely 10 on the first proposal and $20 \times (1 + .7) = 34$ on the second proposal. Thus, the government overpays by 14. If the second proposal is the one selected in the sample, the government finds the .7 padding factor and therefore reduces the first proposal by a factor of $1/(1 + .7) = 10/17$, which results in an underaward of $10 - 10 \times (10/17) = 70/17$, or equivalently an overaward of $-70/17$. Thus, the expected overaward to the contractor is:

(overaward if proposal 1 is chosen)
$\times$ (probability proposal 1 is chosen)
$+$ (overaward if proposal 2 is chosen)
$\times$ (probability proposal 2 is chosen)

$$= 14 \times \frac{10}{44} + \left(-\frac{70}{17} \right) \times \frac{34}{44}$$

$$= \frac{140}{44} - \frac{140}{44} = 0.$$

2. With simple random sampling, proposal 1 is selected with probability equal to .5 and proposal 2 is selected with probability equal to .5. Thus, the expected overaward to the contractor is

$$14 \times .5 + \left(-\frac{70}{17} \right) \times .5 = 7 - \frac{35}{17} = \frac{84}{17}.$$

10. The School of Theatre at Florida State University wants the commencement address to be delivered by this year's winner of the Academy Award for best actress.

Vignette 16

4. (a) There are $4 \times 3 \times 2 \times 1 = 24$ permutations of the numbers 1, 2, 3, 4. They are listed below.

1234	2134	3124	4123
1243	2143	3142	4132
1324	2314	3214	4213
1342	2341	3241	4231
1423	2413	3412	4312
1432	2431	3421	4321

(b) $5 \times 4 \times 3 \times 2 \times 1 = 120$.
(c) $366 \times 365 \times \ldots \times 2 \times 1$.

6. The table of random numbers, Table A of Appendix I, will suffice. Give each faculty member a number by assigning 01 to the faculty member whose surname is first alphabetically, 02 to the faculty member whose surname is second alphabetically, and so forth so that the 90 numbers 01, 02, ..., 90 correspond to the 90 faculty members. Now pick a starting point at random in Table A and record consecutive two-digit numbers (discarding duplicates and 00, 91, 92, . . . , 99) until 10 numbers out of 01, 02, . . . , 90 are selected. The 10 faculty members corresponding to the 10 numbers selected are allowed to attend the special meeting.

7. Suppose in a given year that people with numbers between 01 and 99 are likely to be drafted, whereas those having higher numbers are relatively unlikely to be drafted. It would be possible for June and September to have the same average lottery number, but for June to have 10 numbers between 01 and 99 and September to have no numbers between 01 and 99. This is one disadvantage of using the average lottery number for a month as indicative of the numbers assigned to the days of that month.

Vignette 17

2. The "months to promotion" value for A. B. is $12 \times 18 = 216$. Then in a joint ranking of the 3 females and 24 males the ranks obtained by the females are 24, 25, and 26. The sum of the ranks obtained by the females is 75. Now note that there are only 7 possible joint rankings that can produce a sum as large or larger than the observed sum of 75. They are listed below in terms of the ranks obtained by the females.

Female Ranks	Sum of Female Ranks
27 26 25	78
27 26 24	77
27 26 23	76
27 26 22	75
27 25 24	76
27 25 23	75
26 25 24	75

Since there are 7 joint rankings where the rank sum is 75 or greater, the chance of obtaining a rank sum that is greater than or equal to 75 when the null hypothesis is true is $7/\binom{27}{3}$. From Table B of Appendix I we find $\binom{27}{3} = 2,925$. Thus, the desired probability is $7/2,925 = .0024$. That is, when there are 3 females, 24 males, and the null hypothesis is true, there is only a chance of 24 in 10,000 of observing a female rank sum as large or larger than 75. Hence, we view this large rank sum of 75 as strong evidence supporting the alternative hypothesis that female waiting times to promotion tend to be longer than male waiting times. The evidence is stronger in this hypothetical example than in the real example of Vignette 17 (as is seen by comparing the probability .0024 with the probability .0123 computed in the text, and noting the former is smaller than the latter).

7. One possibility is the **median** of the ranks obtained by the females. For the example in Vignette 17, the median of 24,25 is $(24 + 25)/2 = 24.5$. For the hypothetical example of Problem 2 of Vignette 17, the median is the middle value of 24, 25, 26, namely 25. See Problem 9 of Vignette 22 for further discussion of the median.

8. It is tedious to rank a large number of observations.

Vignette 18

1. Roughly speaking, the rank totals are not close together because the "16" for the "both sexes" category is far from the "30" and "32" totals. We conclude that there is a strong element of agreement.

9. One could define the "best" rum to be the one that has the greatest chance of having the smallest rank total, where the ranks are summed over all the judges. The respective chances for each rum will be unknown parameters, but a reasonable procedure is to pick that rum which has the smallest observed rank total. Applying this procedure to the example of Problem 4, we pick brand C.

Rum:	A	B	C	D	E
Rank total:	37	39	34	38	47

10. Method 1: Have each judge rank all projects from least to greatest, with rank 1 given to the most preferred project, and so on. For each project, obtain its rank total (summed over all the judges). The winner is then the project with the smallest rank total. If several projects are tied by this criterion, have multiple winners.

Method 2: As in 1, have each judge rank all projects from least to greatest. Take the winner to be the project that is most frequently awarded rank 1. If several projects are tied by this criterion, have multiple winners. (This scheme could lead to a large number of winners.)

Vignette 19

6. Two correlation problems: (a) Study the relationship between brightness of plumage and length of life in a species of blackbird. (b) Study the relationship between the distance from the heel of the back foot to the back line of the batter's box and batting average in major league baseball players.

Two regression problems: (a) Control the levels of radiation and study the relationship between radiation and taste aversion (as measured by saccharin consumption) in rats. (b) Control the amounts of information given to workers concerning their success at performing a repetitive industrial task and study the relationship between amount of information and amount of output.

8.   Adding a few points to a scatterplot based on 44 pairs of measurements will not significantly affect the visual indication because the information from the 44 pairs will dominate. However, if the original scatterplot contained only 7 or 8 points (say), then a few additional points would have a proportionately larger effect and could alter the visual indication.

9.   A scatterplot that looks like the letter "V" indicates that the variable represented by the vertical axis first decreases in a straight-line relationship as the variable represented by the horizontal axis increases, and then after a point it increases in a straight-line relationship as the latter increases.

Vignette 20

4.   No. Matching would require that there be enough female and male pharmacists at Lagranze so that in each pair a male and female with similar years of experience, educational background, and so forth, could be matched. It is conceivable that a matched study could be done if other pharmaceutical companies were also used.

5.   Twins obviously agree on the age variable, and are very likely to be similar on other factors that may be relevant to the study.

7.   Retrospective studies are usually faster to perform and less expensive than other types of studies.

Vignette 21

1.   From Table 21.1 we find that for a 58-year-old person, the average number of years left is 19.14. Since Table 21.1 is over 20 years old, this entry is likely to be too low. Improvements in medicine, style of living (including concern with proper exercise and diet), and so on, have extended average remaining life.

3.   The newborn babe is vulnerable to high "infant mortality." Once the child has shown the ability to live for a year, he/she has demonstrated a degree of hardiness. You might say used babies are better than new babies.

5.   81-82.

Vignette 22

1.   7/13.

8.   Order the observed failure times. Starting at 0, begin a horizontal line at a height of 0. At each failure, begin a new horizontal line the height of which is *increased* from the height of its successor by the amount: (number of failures at that given time) $\times$ (1/total number of observations). At the time of the last failure, the height of the horizontal line reaches 1 and remains at 1 thereafter.

10. There were 306 deaths in the interval 41-42, and let's assume that these deaths are uniformly spaced throughout the interval. Then we would estimate that of the 306 who died in the interval, half, namely 153, died by age 41½. Thus, we would estimate the chance of dying by age 41½, given survival up to age 41, as 153/92,785 = .00165.

Vignette 23

4. (a) Pool the reversed values 63.6, 63.5, thereby replacing each by the average (63.6 + 63.5)/2 = 63.55 so that the estimated average heights are now as follows:

1	2	3	4	5	6
63.4	63.55	63.55	63.7	63.8	64.0

These values are now nondecreasing and are consistent with the fact that boys tend to grow taller as they get older.

(b) Yes, a reversal would be less likely. Now each average is based on more observations; thus, each average is a better estimate of the corresponding true population height it is estimating. The improved estimates have a greater chance of having the nondecreasing property that is present in the parameters being estimated.

6. Since system reliability increases as the development program progresses, it is beneficial for the designer to use an estimate that is more likely to reflect this fact. From a mercenary point of view, the contractor can use the higher estimate of current reliability to charge higher prices in the future.

9. A tennis instructor is teaching the serve to a student. There are many individual steps which must be properly coordinated to produce a good serve. These include a proper backswing, a smooth toss, balanced weight transfer, and so on. Each time the student commits an error on one of the steps, the instructor makes a correction, thereby reducing the chance of further errors by the student at that step.

Vignette 24

1. 8/1747.

5. (a) Assume that any of the 31 days in March are equally likely to be M. H.'s birthday. Since 3 of the 31 days received lottery numbers between

1 and 100, the chance is 3/31. (b) Assume that the first 22 days are equally likely. Since 2 of those 22 days received lottery numbers between 1 and 100, the chance is 2/22.

    9. (a) 7/14. (b) Out of the eight women (numbers 1, 6, 8, 9, 10, 12, 13, 14) who selected "women" as most preferred, seven (numbers 1, 8, 9, 10, 12, 13, 14) picked "both sexes" as the second choice. Thus, the desired probability is 7/8.

Vignette 25

    3. We use the same reasoning as in the birthday calculation.

Probability that the six students all have different birthmonths

$$= \frac{12 \times 11 \times 10 \times 9 \times 8 \times 7}{12 \times 12 \times 12 \times 12 \times 12 \times 12} = .223,$$

and

Probability that at least two have the same birthmonth
$$= 1 - .223 = .777.$$

    4. No. For example, in some regions of the country it gets very cold during the winter months, and thus there may be more opportunities for conception during those months. Hence, one may expect more births nine months following the winter months, that is in September, October, and November.

    5. If you make the same assumptions as in the birthday problem, you may use Table 25.1 of Vignette 25 to find the chance to be .654.

Vignette 26

    2. You should bet on a head. You have no assurance that the coin is fair and indeed on the basis of the 100 trials you should estimate the chance of heads on the next trial to be .6.

    5. Some possibilities: (a) Someone is cheating to make the total number of heads equal to the total number of tails. (b) You have observed a very rare event. (c) Some of the questions in this book are loaded.

    7. Assuming that indeed the wheel is well balanced and that successive spins are independent, your answer should be no.

Appendix III
List of Invented Quotations

First Journey Home   Margo Brooks
The Lady and the Statistician   Ingram J. Carswell
Summer Lapse   Mark Hampton
Death Row   Manfred Herst
California Curry   Joyce McShane

Commemoration

To the memory of the 10 goldfish subjected to methylmercury poisoning in Problem 10, Vignette 22.

To the fish who experienced double jeopardy and repeated humiliation in Vignette 14.

To the vast mass of random numbers that never experienced selection, but instead suffered repeated neglect in Vignette 12.

To the enormous numbers of frustrated monkeys who typed year after year in the jungle but never succeeded in recreating even a single line of Milton's immortal epic *Paradise Lost* in Vignette 26.

To the students who were required to share their birthday celebration with one or more students in class in Vignette 25.

To the harassed coins incessantly being flipped, landing on their heads or tails, never experiencing the fulfillment of being traded or the happiness of making new friends in Vignette 26.

To the girls whose beauty and charm were never even seen by their potential suitors, but who were accepted or rejected for marriage solely on the basis of their worldly goods in Vignette 2.

To the millions of alcoholics who live (and die) by the code: If One Is Good, Why Not Have Two? (Vignette 7).

To the persistent system players of roulette who lost all their money and encountered further embarrassment when their requests to have return airplane tickets converted to chips were mercifully denied by the casinos in Vignette 1.

To the dissatisfied major league baseball players who year after year play for noncontending teams and have to watch each World Series on television (Vignette 9).

To the young ladies of the pharmaceutical company who were generously given an enormous amount of experience before promotion, while their male counterparts were rushed into promotion in Vignette 17.

And finally, to the algebraic symbols and Greek letters which were ignored in writing the book, suffering rejection and loss of esteem since they were unable to baffle, bewilder, and impress the reader as usual in a book on statistics.

239

Index